C000203241

Dear
Nici,
Enjoy the bees x
Just being. Be kind to
yourself. Thanks for your
friendship over the years.
Much love
Clare
_x x x _

the Way *to* Bee

the Way *to* Bee
Meditation and the Art of Beekeeping

Mark Magill

LYONS PRESS
Guilford, Connecticut

An imprint of Globe Pequot Press

First Lyons Press edition, 2011

Text copyright © Mark Magill 2010
Layout & Design copyright © Ivy Press Limited 2010

Lyons Press is an imprint of Globe Pequot Press

Library of Congress Cataloging-in-Publication Data
is available on file.

ISBN: 978-0-7627-7365-7

This book was conceived, designed, and produced by

Leaping Hare Press
210 High Street
Lewes, East Sussex
BN7 2NS, UK

Creative Director PETER BRIDGEWATER
Publisher JASON HOOK
Art Director WAYNE BLADES
Senior Editor POLITA ANDERSON
Designer BERNARD HIGTON
Illustrator BELEN GOMEZ

Printed in China
Color Origination by Ivy Press Reprographics

10 9 8 7 6 5 4 3 2 1

CONTENTS

INTRODUCTION

Yes, bees buzz. Any two-year-old will tell you.
But they also hum and murmur and rattle.
Sometimes when you tap the side of the hive in the
dead of winter, they'll make a sound like rustling
leaves. When angry they'll whine like a high-powered
mosquito. But on those warm, sunny afternoons in
spring when they know for certain that winter will
have to wait its turn for another year, the bees
will send up a sound that is like no other.
It's the sound of contentment.

THIS BUSINESS OF "I"

◆

"Do you get stung a lot?" That's usually the first question. "Does it hurt?" That's the second. "Not so much any more," and "you get used to it." It took a while to work out what was wrong with those answers. In the beginning, I mistook getting stung for a mark of distinction. I kept bees. I should expect to get stung. You brush it off and move on. After a season of stings, the venom seems to have less effect. "You get used to it." It even starts to promise a welcome jolt, something like a shot of espresso.

I T TOOK A WHILE to recognize this as foolish pride. And worse. A Greek might call it hubris. Here's one example. On a warm, sunny day in the middle of summer, with nectar flowing from fields of flowers, the bees seem content. With a steady supply of nectar coming in, they don't seem to mind if you open the hive to have a look. This holds true on a sunny day in summer. But it's a different story when it's a balmy day in late October. It was a day like that when I walked up the hill to the bee garden. I saw the bees buzzing around and thought they wouldn't mind if I took a peek. Within the few seconds it took to lift the lid, I had sixteen stings through my jeans. I had neglected to consider that a hard frost had killed all the flowers a couple of weeks before. There would be no more nectar until spring. That meant the bees had no chance of replenishing their stores of honey for at least six months.

Although the weather was deceptively similar to a day in June, the bees' disposition was not. With my lack of insight, I had failed to see this. And the bees paid the price for my ignorance.

Honeybees die when they sting. Unlike wasps, honeybees live solely on pollen and the honey they make from the nectar of flowers. Most wasps are carnivores. They hunt and kill for food. They can sting repeatedly in pursuit of a meal. Stinging for a honeybee, however, is an expensive proposition. They need a pretty good reason to sacrifice themselves.

For honeybees that reason is the defense of their hive. There's little else that will set them off. They are protecting the honey they need to survive the winter. They are protecting their only shelter—their hive—which takes tremendous effort to build. They are protecting their queen, without whom there is no future. And they are protecting their young, whom they nurture through their first weeks of life before they can fend for themselves.

The Greeks Had a Word for It

Hubris is defined as extreme haughtiness or arrogance. Hubris often indicates a loss of touch with reality and overestimating one's own competence or capabilities, especially for people in positions of power. It might seem a stretch to equate the mistakes of a beginning beekeeper with the stuff of tragedy. But standing there in my beekeeper's suit, lording it over the hives, I could say I was assuming a stance that ignored the clear rules of the bees' nature. It didn't feel like arrogance in the moment. I suppose hubris rarely does. But certainly I was overestimating my own competence and capabilities.

It turns out my idea of myself gets in the way of lots of things. I had an *idea* of myself as a beekeeper and was acting accordingly. Instead of looking at what I was doing, I was causing bees to die needlessly and thinking it was normal, even admirable, since I saw myself as a beekeeper. I was getting stung and bees were dying because my idea of myself as a beekeeper was getting in the way of seeing what was really going on. In my ignorance I was doing little more than making thoughtless mistakes. It was certainly nothing to be proud of.

If this sounds vaguely like the state of world politics or the remains of someone's failed relationship, then there's probably a good reason. We sit around scratching our heads along with Robert Burns, wondering why the best-laid

> But, Mousie, thou art no thy lane,
> In proving foresight may be vain;
> The best-laid schemes o' mice an' men
> Gang aft agley,
> An' lea'e us nought but grief an' pain,
> For promis'd joy!
> Still thou art blest, compar'd wi' me
> The present only toucheth thee:
> But, Och! I backward cast my e'e.
> On prospects drear!
> An' forward, tho' I canna see,
> I guess an' fear!
>
> FROM "TO A MOUSE, ON TURNING HER UP IN HER NEST WITH THE PLOUGH," ROBERT BURNS

schemes of mice and men often go awry. The difference between mice and men, according to Burns, is that at least the mice only need to live and experience the moment's tumult. We—blessed with foresight and hindsight—face a more complicated challenge.

What Does Meditation Have to Do with It?

For a beekeeper, the challenge is to avoid causing suffering, for both the beekeeper and the bees. As a beekeeper, there is also a question of accepting responsibility for the welfare of the bees under my care. How do I avoid causing suffering and carry out my responsibilities? I face the same questions as a father, a husband, a son, a citizen. One way would be to understand why our best-laid plans often seem to go wrong. I doubt many people step up to the altar planning to wed in order to enrich some lawyers in a rancorous divorce battle a few years down the road. Yet this still happens as often as not.

Why do my plans go awry and what can I do about it? The simple answer would be to really know what I'm doing instead of imagining I know what I'm doing. Like any task, beekeeping requires attention, observation, and a willingness to learn. From that come knowledge, skill, and accomplishment. What then, stands in the way of that? On one level it can be inattention, a lack of observation, and a closed mind. Are there antidotes to those obstacles?

This is where meditation comes in.

WHAT IS MEDITATION?

◆

The word "meditation" can conjure an image of someone sitting cross-legged with eyes half-closed—a couple of candles and a whiff of incense swirling in the background. Immoveable, tranquil, all thoughts eliminated. It looks good from the outside. But it's what is going on in the mind of the meditator that really matters. The kind of meditation that I'm going to be addressing in this book involves learning, analysis, and concentration.

LEARNING INVOLVES GATHERING the proper information that can help you follow the path you have chosen. Analysis involves examining the information until you are convinced that it is valid and useful. The Buddha encouraged, and even admonished, his followers not to accept his teachings simply out of respect because he was the one teaching them. He told them to test his words "as the wise test gold by burning, cutting, and rubbing it." The process for testing the teachings is analysis supported by experience. Experience, in this case, can come through concentration on the points arrived at through analytical meditation.

Concentration is the ability to focus the mind solely on the subject you have chosen for your meditation. It means that the mind has been trained so that it is no longer jumping around after every distraction, like a monkey in a temple or a wild bull running amok. In fact, in a well-known series of instructional paintings on Zen meditation, which are known as the *Ten Ox-Herding Pictures*, this achievement is referred to as catching and taming the bull.

Meditation relies on mindfulness to ensure that the meditation and the meditator are heading in the right direction, like the meditator riding the powerful but now-tamed bull in the ox-herding pictures. By the right direction I mean overcoming inattention, lack of observation, and a closed mind. It leads to the elimination of the addiction to negative emotions that arise from and are supported by delusion and ignorance. It means putting an end to hubris.

Mindfulness and Analysis

These are two tools that can weaken the grip of negative emotions and overcome delusion. Mindfulness, in this sense then, is the willingness to engage in the repeated practice of being aware of thoughts and actions. It may not always be available in the moment at first, but mindfulness means developing the kind of observational distance that allows me to say "Hey, that was anger," rather than continuing a habitual justification of why anger was the "correct" response to a situation. If I am able to acknowledge what just happened or what is happening, then the second tool—analysis—becomes possible. Analysis enables me to question the validity of justifying anger (or the other negative emotions) and begin to see the foundation of falsehood it is based on. This kind of seeing and knowing weakens its ability to "take" us or to be taken in for too long. Once you start to see the cost of those negative emotions, they lose some of their attraction. In other words, you've got to be able to face it to fix it.

BACK TO THE BEES

Meditation requires a certain kind of effort. But it should also be a welcome and enjoyable effort, with real prospects of reward as well as a sense of accomplishment. When the Buddha sat down under the bodhi tree, he had a definite goal in mind. In fact, he vowed that he would not get up until he had achieved enlightenment. And what did that mean? It meant understanding suffering and what causes suffering. It also meant discovering if it was possible to end suffering and the way to accomplish that.

THIS DID NOT MEAN SITTING with a mind that is completely blank. It meant using extreme concentration and insight to penetrate through all forms of delusion to arrive at the truth. For the Buddha, the goal of his meditation was to achieve freedom—both for himself and for others.

It would be misleading to say I keep bees as a form of meditation. I keep bees because I enjoy working with them. I enjoy observing them. I like the work of harvesting honey and I enjoy being able to give it away to family and friends.

INTRODUCTION

> I hope that you understand what the word "spiritual" really means. It means to search for—to investigate—the true nature of the mind. There's nothing spiritual outside. My rosary isn't spiritual; my robes aren't spiritual. Spiritual means the mind and spiritual people are those who seek its nature.
>
> LAMA THUBTEN YESHE

But it would also be untrue to say I haven't learned from them. It requires effort. Sometimes quite a lot. But it is an enjoyable effort, with enjoyable results when things go well.

A Note

Some of the autobiographical sections in this book are slightly fictionalized out of deference to my farmer friends who have been part of my beekeeping adventures. I distance myself from them by using the adjective "farmer" out of respect. Although I own—or at least continue to pay a mortgage on—some farmland and have had the honor of serving on the county agricultural committee, I know with confidence that I cannot match the level of regard with which I hold these "farmer" friends. So I have changed some names and altered certain aspects of these personal accounts in an attempt to "protect the innocent." I'm sure if my friends should ever happen to read this, they will know exactly whom I am talking about. If so, I'll have to learn to live with the consequences.

15

CHAPTER ONE

SPRING

The ice on the pond grows thinner.
It will not last long. The last patches of snow linger
only in the shadows of the north-facing hills.
The brown-tufted hayfields take on the slightest tinge
of green. One warm day and the tide of winter will
turn for good. Inside the hive, the bees are ready at last
to spring forth from their long confinement. Each day
since the cold began its retreat, the queen has added a
thousand more bees to her ranks in preparation for the
advent of spring. Soon the blossoms will open and the
bees will begin their sweet invasion.

SOMETHING STIRRING

"You figure on bringing us some jars this fall?"

Buddy Dial and his brother Bill ran the farm stand nearby that stocked local honey, including the Dial brothers' own. I would harvest each October after the frost had taken the flowers. I'd bring Buddy a few jars just to keep it interesting. Buddy sold the honey along with his pumpkins and apples and Christmas trees during the months of November and December.

"Too soon to tell," I answered.

At that point I'd lost hives three winters in a row. I had to start up with new bees each spring. It wasn't easy. There were parasites and the hard winters. There were the bears. This was way before the strange new problem of colony collapse disorder.

"I heard Bill Blanchard's packed it in," Buddy was saying.

Bill Blanchard was a commercial beekeeper who kept his hives in the flats down by the river. He supplied the local supermarket with a selection of one- and five-pound jars.

"Between the bears and the bugs, he couldn't keep up," Buddy continued.

The bears had become more fearless over the years. It used to be a rare sight to spot one up on the ridges. My dairy farmer friend Roy Kaiser had worked his fields for forty years and said he never saw a live one. Now they were stealing garbage off the verandahs. They could tear through a bee garden in a night and leave a trail of ruin by morning.

"What's Bill going to do?" I asked.

"Sold off his equipment to an outfit down south. Got a job driving for the feed mill, where his brother-in-law's the foreman," Buddy said.

My main business is writing. The bees were never more than a sideline but, despite the struggles, it was one that I was reluctant to let go.

It wasn't always that rough. The Dial brothers kept about 30 hives. I usually had about 10. Bill Blanchard had close to 300 hives when he was going at it full time.

It used to be the case that a colony would reach a certain size by early summer and you could split them into two hives before they swarmed. And most hives would make it through the winter so that they'd be going full tilt even as the sun was melting the last of the snowdrifts in late March. It was easy to feel a sense of increase. What you might call an embarrassment of riches.

But of course things change. The first time my hives were wiped out, I felt an overwhelming sense of guilt and failure. It was as if I had done something dreadfully wrong and there was no way to atone for it. I was down in the dumps for months over it. My good friend Tom had been keeping bees as long as I had. When he lost his bees the first time, he gave up for a year. It must seem an odd thing to someone who has spent no time with the bees, but the blow of loss comes pretty hard.

Now Blanchard was out. The Dial brothers would quit by next spring after the mites wiped them out. It has been an uphill battle for quite a while now. Colony collapse disorder is just the most recent challenge. There are still bees and beekeepers around, to be sure. But there's little doubt that things are changing.

"Now Ends the Winter ..."

The snowdrifts linger into the end of March in the hills where I keep my bees. That's when the weather finally begins to turn warm enough for the bees to fly in earnest. They make forays now and again on the few warmer days of winter.

Some don't make it back. Strange to say, you can judge the health of the hive by finding the bees scattered in the snow. Their bodies are darker than the snow, so their corpses absorb some of the winter sun's heat and can be found sunken into small depressions in the drifts. I find it a bittersweet ending. Necessary in the scheme of things. But final.

Up to that point in the season, there is little evidence to tell how they are faring. They can suffer from a dysentery-like disease called nosema. You will find the front of the hive covered with small stains if that is the case. The main way to check the hives before the bees start to fly again is by placing your ear against the side of the hive and rapping it with your knuckles. You can hear a healthy hive stirring at your disturbance. Silence is a bad sign.

A Sign of Life

I had checked the hives during the depths of winter. I walked to the side of a hive. I bent over and put my ear to the white wooden box. I rapped it once with my knuckle and heard a rising, rustling hum as the bees stirred to the knock. I went to each of the hives in turn and checked them the same way. Eight still had the sound of life. Two were silent. It would be six weeks before I'd know for sure if they'd made it. But the fact that they had come this far was something.

A BRIEF HISTORY OF BEEKEEPING

◆

The record of honey hunting dates back to the Paleolithic Era. Cave paintings show hunters climbing rickety ladders to the tops of trees or overhanging cliffs to reach into the hives of live bees, risking their necks and the prospect of painful stings to gather honey. The keeping of bees in domestic hives is recorded in Egyptian tombs. People have kept bees in clay cylinders, in hollow logs, in wicker baskets, and in trees with trap doors.

THE ANCIENT ART OF BEELINING was a common practice for millennia, when honey gathering came mainly from wild hives. It is based on the bees' unerring ability to find their way home. In beelining, the honey hunters use a honey-baited box to try and catch some bees.

The hunters release one or two bees at a time and then observe the direction of their flight. The bees will fly in a straight line when returning to their hive. The bee hunters can follow the bees into the woods or wherever else they

> Come now and I'll impart the qualities Jupiter
> himself gave bees, for which reward they followed
> after the melodious sounds and clashing bronze of the
> Curetes, and fed Heaven's king in the Dictean cave.
> They alone hold children in common: own the
> roofs of their city as one: and pass their life under
> the might of the law.
>
> FROM *GEORGICS, BOOK IV,*
> *BEEKEEPING (APICULTURE)*, VIRGIL

suspect there may be a hive of wild bees. Beeliners can also move their beelining box to a different location and then release more bees in order to triangulate their position. Unfortunately, beelining is little practiced nowadays—with the spread of domestically kept hives, it has become more of a sport than an economic enterprise.

Bee Species

There are four races of Western honeybees (*Apis mellifera*) currently considered of economic value, in that they can be kept and managed in sufficient numbers to produce honey and perform crop pollination. These races are the Caucasian, the Carniolan, the dark bee of northern Europe, and, lastly, the Italian. In all, there are twenty-four subspecies of Western honeybee. They are grouped into four categories—the African group, the Near East group, the Central Mediterranean/ Southeastern European group, and the Western Mediterranean/Northwestern European group.

Most of these subspecies of honeybees have adapted to the environment in a particular geographic location as a result of natural selection. In general they differ from one another in terms of coloration, but there is enough variation within each type to make color an unreliable means of distinguishing one type of worker from another.

Beekeepers have identified certain traits that they associate with each type. Aggressiveness, productivity, inclination to swarm, and resistance to disease are some of the qualities beekeepers look for in favoring one type over another.

The proliferation and commercialization of beekeeping and bee breeding since the middle of the nineteenth century has had an impact on natural selection and diversity of honeybees.

The Buckfast Bee

Brother Adam arrived at Buckfast Abbey in England from Germany in 1910 when he was twelve years old. He was soon put to work beside the Abbey's head beekeeper, Brother Columban. At the time, the British black bee was the predominant bee in the British Isles. Italian bees had first been imported in 1859. Aside from the introduction of the Italian bees, no effort had been made to improve on the existing strains of bees through breeding.

The Mite Brings Devastation

Unfortunately shortly after Brother Adam began his bee-keeping, the parasitic acarine mite wiped out thirty of the Abbey's forty-six colonies. All of the bees that died were of the native British black bee variety. The bees that survived the outbreak were all of Italian origin.

Archaeological, biological, and historical evidence had shown that dark European honeybees (*Apis mellifera mellifera*) were the predominant honeybees in the UK from 4,000 years ago up until the nineteenth century.

Between 1916 and 1925, however, the British black bee was rendered virtually extinct in England and Wales by the actions of the acarine mite.

Brother Adam to the Rescue

This devastation impelled Brother Adam to seek a solution. He admired the traits of the British black bee and was unwilling to see them vanish without a struggle. He noted that the Italian bees had survived the plague of mites. He thought it might be possible to introduce the Italian bee's mite-resistant quality into the black bee.

He began a lifelong search for the best strains of bees. He concentrated on countries that had a distinct indigenous race of bees, going mainly to isolated regions where the purity of the native strains had been maintained. He spent nearly fifty years traveling throughout some of the remote sections of Europe, the Middle East, and Africa in search of genetic stock from which he eventually managed to breed a hybrid that became known as the Buckfast bee.

Brother Adam's Paradox

Here's where Robert Burns and his best-laid schemes o' mice an' men come in. Brother Adam was trying to preserve the best strains of bees to breed a honeybee with the most desirable qualities. He did what human beings have been doing since the beginnings of agriculture. He was trying to speed up the process of natural selection by breeding a hybrid.

Nearly all the domestic crops and animals we have today are the result of this intervention. Robert Burns himself was engaged in a form of this practice when he turned up the mouse's nest with his plow.

The consequence of breeding and hybridization, however, is a lack of diversity. Modern agriculture is dominated by a few strains of corn, wheat, or cattle. Ironically, the diversity that Brother Adam was trying so hard to preserve was overcome by his and others' successes.

An Unfortunate Conclusion

As Brother Adam himself put it: "It may not perhaps be generally realized that many races and strains are gradually but surely approaching extinction, due to widespread hybridization. Indeed, according to my findings, this regrettable state of affairs has already progressed so far that the native bee, in its original purity, does not exist any longer in a number of countries —or if still extant, it is only found in remote parts of isolated valleys, shut off from general intercourse. This holds good at least in respect of Western Europe. From the genetical point of view this is a most deplorable development, for many valuable characteristics have been submerged or already completely lost in the welter of indiscriminate hybridization."

THE ARCHITECTURE OF THE HIVE

To begin keeping bees, one needs a place to keep them. That means a beehive and a place to put it. Since 1851 most beekeepers have adopted something called the Langstroth hive; this modern beehive was developed by the Reverend Lorenzo Langstroth in that year and patented shortly afterward.

BORN IN PHILADELPHIA, Pennsylvania, in 1810, Langstroth reportedly had an interest in social insects such as ants from the time he was a youngster. He graduated from Yale University in 1831, and subsequently held a tutorship there from 1834 to 1835. After leaving Yale, he served as pastor of various Congregational churches in Massachusetts. In 1848 he became head teacher of a young ladies' school in Philadelphia. He took up beekeeping, because he was "debarred, to a great extent, by ill health, from the appropriated duties of my profession, and compelled to seek an employment [that] called me as much as possible into the open air."

Before Langstroth

The hives in use prior to Langstroth's design were structures where the bees adhered their comb to the walls and roof of the hive. Harvesting honey involved a serious amount of disruption and destruction to the hive. The bees needed to be driven from the hives with sulfurous smoke and the comb had to be cut from the hive in order to extract the honey. Langstroth was loath to cause this kind of damage and sought a better way.

Bee Space

Langstroth's hive is based on his observation of something called "bee space." Bee space is the $^3/_8$-inch (9.5-mm) space the bees naturally leave to move around between the hive structures they build. Langstroth observed that bees will try to fill in any gaps greater than $^3/_8$ inch (9.5 mm). The bees use wax or propolis for this. Propolis is a sticky resin that bees gather from the sap of trees and use as a kind of cement. For a long time, beekeepers believed the bees used propolis to close up chinks to protect the hives from the weather. However, it is now known that hives benefit from good ventilation, but that bees will use propolis to reinforce the hive and to make it more defendable by sealing up unnecessary entrances.

The Flash of Inspiration

Langstroth realized that if he positioned the cover of the hive exactly ³/₈ inches (9.5 mm) from the top of the frames where the bees would draw their honeycomb, the bees would not attempt to fill the space with wax or propolis. As a result, the cover of the hive could be removed without causing any trouble or damage to the hive.

Langstroth also realized that he could position the frames within the hive in such a way that the bees would build honeycomb on the frames but still leave their bee-space gap between the frames. That meant that the frames of honeycomb could also be removed without damage to the hive or to the bees themselves.

The discovery of this small, simple fact—which had eluded the first 10,000 years of beekeepers—meant that Langstroth was able to develop a hive well suited to the needs of both the bees and the beekeeper. It was a radical departure from the dome-shaped, woven wicker skeps that are still popular today in the depiction of beehives.

Langstroth's Hive

The modern Langstroth hive consists of empty wooden boxes about the size and shape of one section of a file cabinet. The boxes, called hive bodies, have no top or bottom. There is a ledge at the top of the box on which hang nine or ten wooden frames. The top of the frame is ³/₈ inches (9.5 mm) below the

---

Langstroth's Analysis

"… the chief peculiarity in my hive was the facility with which they could be removed without enraging the bees … I could dispense with natural swarming, and yet multiply colonies with greater rapidity and certainty than by the common methods … feeble colonies could be strengthened, and those which had lost their queen furnished with the means of obtaining another. … If I suspected that anything was wrong with a hive, I could quickly ascertain its true condition, and apply the proper remedies."

---

level of a removable cover that sits on top of the box, known as an inner cover. You can stack one box on top of another to expand the size of the hive and accommodate an increasing population. Langstroth's bee space prevails throughout—³⁄₈ inches (9.5 mm) separates the top of the frames from the next set above, and ³⁄₈ inches (9.5 mm) separates each frame from the ones that hang beside it.

A Framework for Success

The frames themselves each hold a sheet of beeswax, called foundation. The sheet is imprinted with the familiar hexagonal pattern of honeycomb. The sheet itself is initially just that, a sheet of wax not much thicker than a heavy sheet of paper. The bees will use the sheet as a base to build their honeycomb.

The bees require two hive bodies for their main hive, called the brood chamber. As the bees increase their stores of honey over the summer, the beekeeper can add half-height boxes of nine or ten frames each, called supers. These are the frames that are harvested in the fall for honey, leaving those that are in the brood chamber undisturbed.

The Self-Assembled Hive

These days, beekeeping suppliers provide kits of pre-cut hive bodies, frames, and foundation based on Langstroth's design. The beekeeper assembles these in preparation for a colony of bees. The equipment is generally made from wood and wax, although plastic is sometimes used in place of this. The wooden hive bodies are painted on the outside to protect them from the weather. The inside of the hive bodies and the frames are left untreated.

Since bees in the wild most often build their hives in hollow trees, wood seems the closest material to what they might use in nature. I certainly prefer wood and wax to plastic when building hives before the new bees arrive.

> So work the honeybees;
> Creatures, by a rule in nature teach
> The art of order to a peopled kingdom.
> FROM *HENRY V*, ACT 1, SCENE 2, SHAKESPEARE

LOCATING THE HIVES

◆

Bees have certain requirements when searching for a new home in the wild. When setting up the artificial version, it's good to keep some of these requirements in mind. Bees need a hive of a certain volume to accommodate their population, which numbers above 50,000. They need room for the queen to lay up to a thousand eggs per day, and they also need room to store the honey and pollen that will sustain them over the winter.

Two LANGSTROTH HIVE BODIES, each holding ten frames, are the right size for a functioning hive. When filled, the hive bodies will hold close to 100 pounds (45 kg) of honey—enough to survive a cold winter, if all else goes well.

The hive also needs an opening small enough for the inhabitants to defend against robber bees and predatory insects. The interior of the hive needs to be sheltered from the elements. Unlike common wasps—which build large, gray papery nests that people often mistake for beehives—the honeybees build only wax honeycomb. They do not create any outer covering. As a result they need a ready-made enclosure for their home. A hollow tree with a knothole for an entrance is perfectly suited to meet these requirements. In a pinch, the honeybees will locate their hives inside the walls of houses, in empty barrels, or in other enclosures with the right volume and a defendable entrance.

A Sheltered Spot

When situating hives in a bee garden, there are several other aspects a beekeeper can consider. It is good to locate the hives where they are sheltered from the prevailing winter winds. An eastern exposure to the rising sun will help warm the hives after a chilly night and rouse the bees earlier than their shadowed neighbors. In the colder climes of the northern hemisphere, an exposure to the southern winter sun also helps warm the hives.

REVEREND LANGSTROTH SAVES HIMSELF

Langstroth referred to himself as being in "ill health" and most references interpret this as "bouts of severe depression." I have spent some time looking for the evidentiary text where "ill health" becomes "bouts of severe depression," but I have been unable to find it.

I PRESUME THAT SOME OF THOSE who have used the phrase "bouts of severe depression" have had the same difficulty finding the original source for this diagnosis. The temptation—depending, of course, on whether the elusive source is indeed accurate—is to equate Langstroth's beekeeping with a palliative for this condition. I was swayed by the same temptation in giving this section its title—"Reverend Langstroth saves himself."

But in my fruitless search for the evidence where ill health was admitted, diagnosed, or proven to be depression, I realized Langstroth's employment in the open air represented the reason that I admire practical investigators like Langstroth and spiritual seekers such as the Buddha. In the most straightforward terms, they were all looking for the truth.

The Path to Success

Their efforts were essentially a practice that provides the theme for this book:

- **Motivation**
- **Observation**
- **Action**
- **Results**

Beekeeping may not be a cure for depression. But motivation, observation, and action may result in less suffering. Langstroth found that by understanding the needs of the bees, he was able to find a method that served both bee and beekeeper. Through careful observation, he discovered the importance of bee space, and he was able to act on his observations to devise a system of hive construction that employed the principles he observed.

The bees and I benefited from Langstroth's findings just yesterday afternoon. I was able to open the lid of a recently arrived hive, lift out a frame, identify the queen, determine

Meditation is meant for the removal of ignorance …

Concentration and all other practices are meant for recognizing the absence, i.e., non-existence of ignorance. No one can deny his own being. Being is knowledge, i.e., awareness. That awareness implies absence of ignorance.

Therefore everyone naturally admits non-existence of ignorance. And yet why should he suffer? Because he thinks he is this or that. That is wrong. "I am" alone is; and not "I am so and so," or "I am such and such." When existence is absolute, it is right; when it is particularized, it is wrong.

That is the whole truth.

FROM *TALKS WITH SRI RAMANA MAHARSHI,*
SRI RAMANASHRAMAM, 1955

if she was laying eggs and then replace the frame within a matter of minutes. There were no enraged bees, no stings, and I ascertained quickly what I needed to know—that the new queen was alive and well.

In Pursuit of the Truth

I can't say for certain whether Langstroth's ill health was in fact depression. In the end that's more a matter of concern for Reverend Langstroth than for me. What I do see throughout Langstroth's writings is a very clear sense of purpose and satisfaction in his pursuit of the truth.

> Whenever you would unseal their noble home, and
> the honey they keep in store, first bathe the entrance,
> moistening it with a draft of water, and follow it with
> smoke held out in your hand. Their anger knows no
> bounds, and when hurt, they suck venom into their stings,
> and leave their hidden lances fixed in the vein, laying down
> their lives in the wound they make.
>
> FROM *GEORGICS, BOOK IV,*
> *BEEKEEPING (APICULTURE)*, VIRGIL

I don't think anyone would claim that beekeeping in itself is a cure for clinical depression. But when Buddha took his seat beneath the bodhi tree, his unshakeable motivation was to understand the causes and cure for suffering. His goal was to find the truth, then act on it. A cure may lie in that.

I think the search for truth is at the heart of both spirituality and science. Healthy doubt is essential to exploration. Without the kind of positive doubt that is the basis of enquiry, the necessary questions on the path to realization cannot arise. Is this true? Does it tally with my experience? Can my conclusions stand up to rigorous analysis?

Rigid dogma and closed-mindedness are antithetical to the process of realization. The lack of enquiry makes anything but blind faith nearly impossible. Faith in one's convictions is important. But it's better if it's intelligent faith, not blind.

DEVELOPING THE DESIRE TO MEDITATE

◆

*The biggest obstacle to meditation is not a sleepy or distracted mind;
it is laziness. This comes in many forms. There is the regular kind of
laziness, with no energy or motivation. But busy laziness seems to be
our most prevalent modern version. Can't sit still. Not enough time.
Too much to do. Everything takes priority.*

THE PHYSICAL AND MENTAL JOY that develops with the
perfection of meditation can be a direct antidote to the
laziness that stands in the way. But since we can't experience
this antidote without also making a determined effort to
perfect concentrated meditation, it seems like a conundrum.

It's difficult to be able to practice with enthusiasm if I have
no experience of the results. I need to develop a desire to
practice in order to practice with enthusiasm. Desire, in turn,
depends on having confidence in the outcome and its value.
I need to know that if I put my time and energy into this kind
of practice, it's going to prove worthwhile. If I don't see this,
there is not much chance that I will have the enthusiasm and
perseverance to pursue it.

Why Meditate?

To pursue this effort I need to understand the potential
benefits of meditation. Traditional teachings on concentrated
meditation list six main benefits:

• **You will experience physical and mental joy with the perfection of concentrated meditation.** Mind and body will achieve a state of suppleness as a result of the ability to concentrate without effort.

• **You will gain control over your mind.** Like the meditator in the *Ten Ox-Herding Pictures*, you will have tamed your powerful mind and can now direct it to useful purpose instead of constantly chasing after it, like the ox herder chasing the wild bull through brush and brambles.

• **You will be able to focus your mind on positive activity.** Because of your ability to focus, you will see the undesirability of negative thoughts and actions before they have a chance to get a hold of you.

• **Your mental powers and abilities will be increased.** People who have concentrated on mastering specific skills in sports or other activities use the power to focus to give them a great advantage in their fields.

• **Your sleep can be transformed into meditative practice.**

• **You will develop the ability to practice *vipassana*— or special insight.** This will enable you to cut the root of delusion and negative action at its source.

These benefits do not depend on any particular philosophy or religion. They are part of the capability of the human mind, which can be developed through practice and attention.

INTRODUCING THE BEES

◆

Langstroth's search for the truth ended up benefiting both him and the bees. Many other bees and beekeepers have benefited from his insights since then. It was the combination of Langstroth's strong motivation not to cause harm, his personal need to find a practice that would ease his "ill health," and his willingness to carefully observe reality that led to his discovery. These qualities apply equally well to developing the ability to meditate. Strong motivation provides enthusiasm. Powerful need provides patience. A willingness to observe reality provides concentration and wisdom. These qualities probably wouldn't hurt in becoming a better beekeeper, either.

UNLESS YOU'RE ABLE TO CAPTURE A WILD SWARM—which could be risky as they won't be certified disease-free—the principal way to get new bees is as a gift from a friendly beekeeper or to order them from a specialist bee breeder. The American Beekeeping Federation recommends that you join a state or local beekeeping organization in order to gain experience in handling bee colonies and advice on where to obtain bees. You can also take a course before becoming a beekeeper.

Bee breeders usually ship the bees in a special transportation box that contains nucleus frames with a brood, food, bees, and a laying queen. These are transferred to your hive. Or the bees may be sold as a "package," loose in a special traveling box, with the queen in a separate cage.

Welcoming the Queen

If you purchase a "package," when the bees arrive, spray the sides of the screen with sugar water to feed them after their trip. You should have already prepared the empty new hives for their arrival. The first step is then to remove the queen cage, while trying to keep the rest of the bees from flying out. The sugar water helps, since bees can't fly when wet.

Remove the tiny cork and poke a hole through the sugar candy so the queen's attendants and the bees outside can chew the rest of the way through and release her. Hang the cage with the hole at the top in case an attendant bee should die and block the hole before the queen can escape.

Pouring in the Bees

Once this is done, open the box of bees and literally pour them in. The bees are in a swarm mode, ready to build a new hive. As such they engage in something called bridging or festooning, where they cling to one another. This helps them when they're hanging from a tree branch as a swarm while waiting for scout bees to find a new location for the hive. It also helps when they're bridging the gaps while building comb in their new hive. When bridging, the bees seem to pour from their shipping container like slow-moving molasses.

... the bees are in a swarm mode, ready to build a new hive.

Settling In

With the bees in their new home, it's time to close the cover and leave them quietly alone. Unless you have some leftover frames of honey, you will need to ensure that you feed them sugar water to keep them going until they've had time to start making honey for themselves.

You need to check after a few days to make sure the queen has been released from her cage. About a week or so after that, you need to check to see if she has begun laying eggs. After that it's up to the bees.

ABOUT THE QUEEN

Each hive has only one queen. Every female worker egg laid by the queen has the potential to become a queen. A queen begins as an ordinary female worker egg deposited by the previous queen in a specially shaped queen cell roughly the size and shape of a peanut shell, or in a regular cell that is then built out by the workers into an emergency supercedure cell in the event of the sudden loss of a queen.

THE QUANTITY AND QUALITY of the food fed to the worker larvae by nurse bees determine whether the female egg will be transformed into a queen. The food fed to developing bees by worker bees during their stint as nurses is largely produced in the hypopharyngeal glands along the side of the worker's head and mandibular glands at the base of the jaw. The brood food produced by these two glands differs somewhat. The results produced by this difference are dramatic.

The food fed to potential queens is called royal jelly, which is rich in products of the mandibular gland. Larvae destined to become workers are fed more hypopharyngeal-gland products as well as more honey and pollen during their last days of larval development.

Queen or Worker?

The developing larvae have the potential to mature into either workers or queens for the first three days of larval growth. But by the fourth day of their development, the larvae are committed to one path or the other.

Since the eggs hatch after three days and the larvae can be transformed into a queen up to three days after they hatch from their eggs, it means the hive has only a six-day window to replace their queen if she should die unexpectedly. Without a queen, the hive is doomed because the workers will not be replaced as they die off.

A Queenless Hive

In a hive without a queen, the female workers will resort to laying eggs themselves, but to no avail. Since the workers have never mated, the eggs they lay remain unfertilized and all will become male drones as only fertilized eggs become females.

The new male drones will be useless to the survival of the colony. The queen only deposits one egg per cell. The workers will deposit two or three eggs in a cell. This is a sure sign of a queenless hive, where the normal order has been overturned.

The larger drone larvae will be unable to metamorphose properly when they are crowded together in a smaller female cell. They either die or emerge in a dwarfed condition. A queenless hive appears agitated and without direction. The honeybees no longer receive from the queen the pheromones or chemical signals that govern the hive. In a normal hive, the bees fly in and out with a purpose. You can see them clambering into the entrance with the loads of pollen and nectar. A queenless hive seems to sound different, with a disordered buzzing when opened rather than a steady hum.

A Queen's Role

The differences between the queen and the workers are striking. The queen is not really that much larger. The main visual difference is in the length of her abdomen, which is the third, rear section of the bee's body. It is perhaps twice as long as a worker's abdomen. But there are other differences. The time of development from egg to emergence is sixteen days for the queen compared to twenty-one days for the workers. A worker will live for about six weeks in summer— when it literally seems to wear itself out—and perhaps twenty weeks over the winter. A queen can live from two to five years.

The queen's major task is egg laying. The workers feed the adult queen brood food and possibly some honey to provide the material for egg production. A queen can lay between 175,000 and 200,000 eggs a year. As such she is the sole mother for many hundreds of thousands of bees over her lifetime.

The queen's other important role is to produce the pheromones that govern and organize many of the colony's functions.

The Anatomy of the Honeycomb

The queen usually begins laying eggs near the center of the hive and then works her way outward as the honeycomb's cells are filled up. She deposits one egg per cell. The queen lays female eggs into the regular-sized cells and the bees produced will become workers. The same-sized cells are used for raising the brood of bees as for storing honey and pollen.

The contents of the honeycomb cells are arranged with the brood in the center, surrounded by a ring of cells filled with pollen, then there is a large outer ring of honey, stored in cells with a protective wax capping.

One final ring of cells is on the periphery of the comb. These cells are about 10 percent larger than the regular cells and they represent 10 percent or fewer of the cells in the hive. This is where the queen lays the eggs that will become the drones or male bees, which represent 10 percent or fewer of the population. The rest of the bees are female—this may be as many as 50,000 per hive.

A Bee is Born

The eggs hatch into tiny larvae after three days. Nurse bees feed the young larvae honey and pollen mixed with hormones from glands on the nurse bee. The honey provides the carbohydrates and the pollen provides the protein the larvae need to grow and metamorphose into bees. The young remain as larvae for six days, after which the nurse bees cover the cells with a porous wax cap. The new bees transform and chew their way out of the cells after another twelve days, making a total of twenty-one days from egg to fully formed bee.

Drones

The workers play a variety of roles in the hive. As only females make honey, the drones have only one role (as far as is known): to mate with the queen. Most die before mating, since a mating queen is a relative rarity and there are many more drones than queens produced in a hive.

A queen only goes on a mating flight at one point in her life. In a hive that has never developed a population large enough to swarm, the opportunity, therefore, for the drones to mate could happen only once in five years. Those few drones that manage to mate with a queen die in the process, since the drone's abdomen bursts during mating.

A queen only goes on a mating flight at one point in her life.

FIRST FLIGHT

A few weeks have gone by since you introduced your consignment of bees to their new home, and the first of the queen's offspring are hatching. A new bee begins its emergence by chewing through the wax capping on the honeycomb cell where it has spent its first twenty-one days. It takes the bee fifteen or twenty minutes of effort before it has finally chewed away the seal. Then it must slowly climb from the cell. During its metamorphosis, the bee has changed from a tiny larva to a fully formed bee.

B UT THE BEE now barely fits in the cell and so the effort to emerge seems arduous. It pauses to rest before finally climbing from the cell. No bee stops to greet it. Instead the workers hustle by, going about their business. The new bee pauses for a moment and then scuttles off to join the flow.

Each new worker will go through a series of roles during its life in the hive. At least 90 percent of the bees are female workers. Before the end of winter, the ratio is closer to 100 percent. As we have seen, male drones perform no work except to mate with a new queen. So after the frost has killed the flowers and cut off the nectar supply, the female workers can be seen dragging the resisting drones to the edge of the hive and then pushing them out into the cold. With no new resources available until spring, the workers conserve honey over the winter by eliminating the drones.

Play Flight

When a bee is about one week old, it takes a brief orientation flight in front of the hive and circles nearby. This instinctual "play flight" enables the bee to learn the hive's appearance and location. The young bees navigate close to the hive by using landmarks and hive appearance. They navigate to and from the hive by using the position and direction of polarized light from the sun. This play flight is most evident on warm, sunny afternoons after several days of rain, when numerous young bees depart the hive together for their orientation. After a few minutes, they return to take up their duties inside.

STAGES OF THE WORKER

The worker's tasks take place in two stages. The bee begins with tasks inside the hive and, with age, transitions to work outside the hive. The new bee's first job is to serve in cell cleaning and capping. This is followed by brood and queen tending. As the bee develops, it then takes on the work of comb building, housekeeping, and transferring nectar and pollen as needed.

THE BEE'S OUTSIDE TASKS begin with ventilating the hive. The watery nectar needs to be evaporated until nearly all of the water content is gone, transforming it to honey. The bees accomplish this by standing at the hive entrance on hot,

Seeing Clearly

Meditation and science both rely on accurate observation. Accurate observation depends, first of all, on identifying and dispelling prejudices, assumptions, and misconceptions that stand in the way of seeing clearly. Clearing these obstacles demands a certain measure of self-knowledge, the kind that is rarely attained overnight. It, too, relies on observation and analysis. This calls for patient dedication to the process of seeing more clearly. It's no surprise that we learn to see by stages. Keener understanding of how we see increases our insight into what we see.

One night on their way home, a group of friends found Mullah Nasruddin—the Sufi wise man and fool—crawling around on his hands and knees under a lamp post, looking intently at the ground as he made his way around.

"What are you searching for, Mullah?" one of his friends asked him.

"I've lost the key to my house," he replied.

The friends joined in to help him look. But after searching in vain, one of the friends thought to ask Nasruddin where he had lost the key in the first place.

"In my house," Nasruddin replied.

"Then why are you looking for it under the lamp post?" the friend asked.

"More light," said Nasruddin.

When Zen master Kyudo Roshi (1927–2007) first heard this story, his instant response was "Searching is the key."

hazy days and fanning with their wings to increase the evaporative airflow through the hive. After their ventilation service is fulfilled, the bee transitions to guard duty, protecting the entrance of the hive and challenging intruders.

Finally the bee becomes a forager, flying distances as far as 3 miles (5 km) from the hive in search of nectar and pollen. The workers can then adjust to the needs of the hive by resuming former roles or advancing to outside duties if required.

LOCATING THE MEDITATOR

There are two kinds of meditation—analytical and concentrated— sometimes called vipassana *and* shamatha, *or insight and calm abiding. Some traditional teachings list six requirements for taming the mind and developing single-pointed concentration. These prerequisites can be taken literally, of course. But these instructions are 2,600 years old and may seem archaic in some regard. Meditation primarily involves training the mind, so these physical instructions can also apply to one's attitude or state of mind. Finding an appropriate place to meditate, however, is fundamental.*

I F YOU'RE KEEPING BEES, you need a place to keep them. The ideal apiary has certain requirements. In order to meditate you need a place to meditate. It also has certain requirements. In the case of meditation, the proper place is traditionally said to have five qualities. These qualities can be taken both literally and figuratively. It is the mental and emotional aspect that is most relevant.

Good Supplies

Traditionally it is said that your meditation should take place in a location where you can find food and other necessities without undue difficulty. If you have to spend all your time worrying about your next meal, it will be difficult to concentrate.

No Danger

In ancient times meditators were instructed to choose a place where they were safe from the dangers of wild animals and hostile individuals. Many of us are able to dwell in places free from deadly animals and enemy attacks. But the real dangers for the meditator are the addiction to negative emotions and the control of the self-destructive ego.

A Healthy Place

Although most of us may be free from the attacks of wild tigers, we face environmental threats, both inner and outer, where things are out of balance. The outer threats are fairly well known thanks to our growing awareness of toxic environments, air and water pollution, and other contaminants such as mold. Yet inner health also requires balance, both physically and mentally. Balance is one of the key requirements of concentrated meditation, where the chief obstacles are a sinking and wandering mind. Rather than being too dull or too tense, meditative balance requires the opposite of sinking and wandering, which means relaxed attentiveness. The Buddha said that the meditator should be like a properly tuned string on a musical instrument, neither too tight nor too loose.

Quietness

Quiet is a definite help in meditation. However, while outer silence can reduce the external distractions, it may make the inner distractions seem more prominent. Concentrated meditation is sometimes called calm abiding, which implies the stillness of silence. An accomplished meditator carries this calmness in the midst of the bustling world. The last of the *Ten Ox-Herding Pictures* is known as *The Old Man in the Marketplace*. This old man walks through the noisy market with "bliss-bestowing hands" and everyone he looks upon becomes enlightened. A quiet place also means refraining from idle chit-chat, which nowadays includes the "noisy" hours spent online. You may notice that this mental chit-chat can also take place quite well even with only one person in the room.

Good Friends

Traditional teachings say that a meditator should associate with virtuous friends. That generally means people who are supportive of your efforts, rather than those who oppose or scoff at what you are trying to accomplish as a meditator. Friends, family members, and others can have a tremendous influence on us and on our progress. Most of us have had the feeling of being influenced by others, either helpfully or harmfully. We can benefit from helpful influence just as we can suffer from harmful influence.

When we recognize and respect the good qualities and accomplishments of others and try to emulate them, we are benefiting from their influence. This includes those good teachers whom we acknowledge and respect. For those who want to learn to meditate, it is important to have good role models as well as good friends. We can use our virtuous friends as well as our role models as a guide and influence in following the meditator's path.

If I behave in the same way as the childish,

I shall certainly proceed to lower realms.

And if I am led there by those unequal (to the Noble Ones),

What is the use of entrusting myself to the childish?

One moment they are friends,

And in the next instant they become enemies.

Since they become angry even in joyful situations,

It is difficult to please ordinary people.

They are angry when something of benefit is said,

And they also turn me away from what is beneficial.

If I do not listen to what they say,

They become angry and hence proceed to the lower realms.

They are envious of superiors, competitive with equals,

Arrogant toward inferiors, conceited when praised,

And if anything unpleasant is said they become angry;

Never is any benefit derived from the childish.

Through associating with the childish,

There will certainly ensue unwholesomeness

Such as praising myself and belittling others

And discussing the joys of cyclic existence.

Devoting myself to others in this way

Will bring about nothing but misfortune,

Because they will not benefit me

And I shall not benefit them.

I should flee far away from childish people.

When they are encountered, though, I should please

Them by being happy.

I should behave well merely out of courtesy,

But not become greatly familiar.

FROM *GUIDE TO THE BODHISATTVA'S WAY OF LIFE*,
CHAPTER 8, VERSES 9–15, SHANTIDEVA

The important thing to remember is that we are trying to develop concentrated meditation, which can help reduce the habit of negative emotions. If you're like me, you can probably use all the help you can get. Negative "childish"

52

friends usually take us in the opposite direction, increasing our "skill" with negative emotions. Better not to be spending our efforts at cross purposes.

Having Few Needs

In developing the ability to practice concentrated meditation, you might find your mind being pushed and pulled by thoughts of needs and wants, both real and imaginary. Take a look around your home. Imagine having to pack up and move everything at a moment's notice. Then imagine having to

---

Tanzen and Ekido were once traveling together down a muddy road. A heavy rain was still falling. Coming around a bend, they met a lovely girl in a silk kimono and sash, unable to cross the intersection.

"Come on, girl," said Tanzen at once. Lifting her in his arms, he carried her over mud.

Ekido did not speak again until that night when they reached a lodging temple. Then he no longer could restrain himself. "We monks don't go near females," he told Tanzen, "especially not young and lovely ones. It is dangerous. Why did you do that?"

"I left the girl at the other side of the road," said Tanzen. "Are you still carrying her?"

FROM *ZEN FLESH, ZEN BONES*, PAUL REPS, 1961

---

discard everything but a few necessary items. What would you choose? What qualifies as necessary? Someday we will really have to go once and for all, leaving all our possessions behind. We won't be able to take anything with us. Yet we carry so much in our minds right now.

In meditation, we can see how we are drawn to all sorts of new "possessions." The best course is to try to acknowledge a few simple needs while practicing meditation: "Why take you thought for raiment? Consider the lilies of the field, how they grow: they toil not, neither do they spin." (Matthew 6:28).

Being Content

According to the Buddha, the lack of contentment is our chief source of suffering. In depictions of the meditating Buddha his right hand touches the earth, in the gesture indicating meditative realization, which represents the eradication of sickness and suffering—and the roots of delusion—through the recognition of absolute truth.

The fundamental cause of suffering is a lack of contentment and the addictive quality of the negative emotions such as jealousy, obsession, and anger. To symbolize the need for contentment, he holds a begging bowl in his left hand.

Lack of contentment can destabilize a meditator's attempts at achieving concentrated meditation. The greener grass on the other side of the fence will be a constant temptation. If you are fortunate to have a place to live, or car to drive, or

> "I believe that the purpose of life is to be happy. From the moment of birth, every human being wants happiness and does not want suffering. Neither social conditioning nor education nor ideology affect this. From the very core of our being, we simply desire contentment. I don't know whether the universe, with countless galaxies, stars, and planets, has a deeper meaning or not, but at the very least, it is clear that we humans who live on this earth face the task of making a happy life for ourselves. Therefore, it is important to discover what will bring about the greatest degree of happiness."
>
> FROM "COMPASSION AND THE INDIVIDUAL,"
> H.H. THE FOURTEENTH DALAI LAMA, 1991

cushion on which to meditate, appreciate it and be satisfied. We use dissatisfaction to constantly drive ourselves to achieve and attain more. From a practical business or career standpoint this seems like a good idea. But dissatisfaction also has a negative, undermining aspect that results in bitterness, enmity, jealousy, and even rage. It would be difficult to make a positive case for any of those outcomes.

The problem is that the pernicious ego uses dissatisfaction as a goad to drive us. This aspect of the ego never seems to be satisfied. Mythology is full of stories of the calamities and misfortunes that unbridled greed—which is really a manifestation of dissatisfaction—can rain down on the greedy.

One of my teachers, Gelek Rimpoche, had a student who owned a Rolls-Royce. They were speaking about the subject of appreciation and contentment. The student asked Rimpoche if there was any problem with owning a Rolls-Royce. Rimpoche said owning a Rolls-Royce was fine as long as the fellow drove the Rolls. When the Rolls-Royce started driving him, that would be a problem.

Retreating from the Demands of Society

Traditional teachings mention retreating from activities such as buying, selling, and idle gossip. For us the push and pull of society's demands represent a kind of busy laziness that distracts us from pursuing the kind of development that can make a real difference in how we manage ourselves and our relationships to society. The ideal is closer to the *Old Man in the Marketplace* with his bliss-bestowing hands, instead of the scattered, multitasking approach that seems to have become a priority. Meditating with my iPhone at my side is probably a pretty good indicator of my priorities at the moment.

Having Pure Ethics

In the traditional monastic sense, having pure ethics means protecting your vows and commitments. What it means in our case is avoiding negative harmful actions. The way to avoid them is first to recognize their negative harmful nature. Lying, stealing, and killing are considered negative in most cultures. But even then people make huge exceptions when it comes to someone they consider the enemy. You can see how someone might be able to develop tremendous concentration when it comes to thinking about how to destroy his or her enemy. But this is a case where the meditative power we are all capable of developing has been completely hijacked by the negative emotions. We need to develop an awareness and aversion to this kind of hijacking attempt if we are interested in developing the more positive aspects of ourselves. If not, there are plenty of hate groups ready to welcome us with open arms, as long as we don't make the mistake of crossing them.

Lessening the Thoughts of Desire

The two main obstacles to actual concentrated meditation are a sinking and wandering mind. There are many levels of sinking and wandering, but sinking mind in the gross sense means falling asleep. Wandering mind is characterized by excitable thoughts that chase after one another like wild monkeys. These thoughts seem propelled by desires and attractions or slights and aversions. The wandering mind needs to be weaned from the distractions of desires and trained to resist the tumult of discursive thoughts. All of these traditional prerequisites to meditation have a bearing on cultivating an attitude that is conducive to meditation. They go a long way toward developing a steadiness of purpose and quelling obsessive thinking.

PREPARING FOR THE SWARM

———————◆———————

Bees multiply their colonies through swarming. It is their form of reproduction on the colony level. Most swarming occurs in mid-spring, so that the bees will have time to find and construct a new home and fill it with honey before the onset of winter. When the population of the hive increases to a certain point, the bees begin making preparations to swarm.

As THE SIZE OF THE POPULATION increases, the transmission of a swarm-inhibiting pheromone, bee to bee, from the queen is reduced because of the number of bees that need to have contact with the pheromone. As a result of this, the workers begin constructing six or more queen cells in which to rear their potential new queens.

Shortly before the swarm departs, the bees gorge themselves with honey that they will later transform to wax to build their new hive. When they are ready, the swarm leaves the hive with the old queen. About 16,000 workers are in a typical swarm, yet the number in the swarm can be as high as 30,000 bees in a strong hive.

A Fight to the Death

The swarm flies off in a loose spherical cloud some 40 or 50 feet (12 or 15 m) across. It can be heard from hundreds of miles away. Once the swarming is completed, the newly emerged queens fight to the death so that only one queen remains in the hive. The queen is the only female that can sting without dying. The drones have no stinger.

The Democracy of the Swarm

Meanwhile the swarm has landed on a tree branch or other convenient perch. They cling together in a mass, with the old queen protected somewhere in their midst. Sometimes a swarm will take off and then have to return to the hive if they discover that their queen has not yet accompanied them. This can happen several times over an afternoon until the swarm finally departs for good.

The swarm will sometimes move from one tree to another before settling in to wait for their scouts to discover a new location for the colony to build their hive.

House Hunting

As the swarm clings to the branch, scouts go out in search of a suitable place. The scouts examine the size of the hive entrance and whether it is sheltered, damp, or dry. They are able to calculate if the volume of the potential home is the right size by walking around the interior. The scouts then return to the swarm. They land on the outside of the swarm and perform a waggle dance to indicate the direction and distance to their nomination for their new location. Other scouts are reporting back with their own nominations.

The scouts will continue lobbying for their candidates until the nominations are narrowed down to a few. They will then inspect the remaining locations and report back.

Putting It to the Vote

If they reach a deadlock in their voting, sometimes half the bees will go to one location and the others will fly to the second. However, they will quickly return and begin voting again. If they cannot arrive at a consensus, the swarm may begin building their comb on the branch where they rest. This is fatal, since the hive will be fully exposed to the elements, with no way to defend it or keep it warm.

If enough scouts report favorably on the same location and agreement is reached, they will begin a whirring dance as they bore their way into the cluster. This alerts the swarm that it is time to depart. A loud buzzing will begin.

A few moments later the first few bees will depart. They are quickly followed by the entire swarm. The bees fly to the new location and pour in through the entrance. Within hours after the swarm has entered, guards have been set and the construction of the new comb has begun.

Mating Flight

The new queen emerges from the hive for her mating flight between five and ten days after the old queen has left the hive with the swarm. This is the only time she will leave the hive until, if all goes well, it is her turn to swarm some time in the future. The queen mates with between seven and seventeen drones during her flight. The drones die during mating. After her flight, if she survives and is not attacked by a bird or other predator, the queen returns to the hive to begin her lifelong business of creating her progeny.

Within hours after the swarm has entered, guards have been set and the construction of the new comb has begun.

CHAPTER TWO

SUMMER

If spring is a time of fertility and fall a time of fruition, summer is one of increase. But even in the midst of increase, there are signs of diminution. From the first day of summer, the days will begin to grow shorter. For the bees this is the season when all the work must be done to prepare for the winter that lies ahead. There are fields of nectar-bearing flowers to harvest while the blooms last. Honey must be stored to feed the hive for the rest of the year. It is a cycle with change and impermanence at its center.

CHASING THE SWARM

As an aspiring writer, I had written a script that turned out to be just what the organizers were looking for at a famous summer camp for writers, high in the mountains of Utah. They sent me a ticket and a check, and told me I'd become one of the chosen few. While I was at summer camp, a niece of Samuel Goldwyn stepped in with the cash to send me to that feeding frenzy called Hollywood. Lots of lunches with movie stars, agents, co-executive producers, the whole suntanned food chain. I took the money and bought a house in the country. I got to know Roy and Charlene Kaiser after I'd moved out there.

I spent most of the first year gutting and rebuilding the house, which was set in dairy-farming country outside of town. The area was a mixture of woods, pastures, and hayfields set in a rolling land-scape of hills. The soil was poor and farming had always been difficult. More easily worked land lay to the south across the river.

It was a very peaceful house. There was something about the scale of the rooms, the wide plank floors, the wood trim of the windows, and the constantly changing light that gave me a feeling of solace. The porch looked out over a valley and a series of ridges to the west. I could see the weather coming and I could watch the course of the setting sun, from the winter to the summer solstice as it made its way north across the distant hills. The verandah was a good place to follow the seasons. It was windy in winter when the snow would fly and the temperature dropped to well below zero. In summer rolling clouds of thunder swept in from the west. I would step out there at all hours to follow the changing light or the rising of the stars.

I got to know my neighbors. I joined the local fire department. And I started keeping bees. My friend Roy Kaiser told me about how he used to help his father in the bee garden. The last few winters had been hard for Roy. All the aches and pains had started to add up and climbing the hill to the bees was a chore for him. The last years had been hard on the bees as well.

The bees would end the season strong enough, but little by little, their numbers would dwindle over the winter. Come spring, when the first warm days should have seen a cloud of bees buzzing around the hive entrance, there'd be nothing but unhealthy silence.

Roy had farmed all his life and he was no stranger to nature's setbacks, yet he found the dying bees a daunting sight. The bees were the most industrious creatures he knew and here they were, laid low.

There were precautions to take and Roy took them. He wrapped the hives in tarpaper and stacked bales of mulch hay around them to keep out the cold. He laid on menthol to deter the mites that were preying on the bees the last few winters. He tracked down abandoned hives and brought them back for breeding, with the belief that the bees that survived in the wild had built up resistance.

Roy took on all his work with that same thoroughness. The years had slowed him down, it's true, but his pace was steady. He had grown up on the farm. He had spent his whole life working the sloping hayfields and pastures. He had seen the changes, great and small. The fading paint on the side of Roy's barn read Kaiser & Son, though his son had long since moved to Florida with never a thought of farming. The apple trees behind the house had lost limbs to wind and time. The abandoned fields across the road had slowly turned back to brush and woodland.

Roy had worked forty hives back when things were rolling. Now he was down to six and that was a struggle. He'd lost his hives three winters in a row, starting up with new bees each spring. It wasn't easy. A spring frost could kill off the first blossoms. A rainy summer could make it difficult for the bees to fly and would set back the flowers that needed warm, sunny days.

Roy's main business was dairy farming. But his family had been keeping bees on their hill for three generations. He had learned the art from his father. Albert Kaiser had been a hard man who had never done better than a threadbare existence at farming. It was one business where you found out early that nature held the upper hand. Some men learned to bend with the wind and others railed against it. Albert was one who shook his fist at the sky. Each fall, he made some applejack and nursed his bitter mood all winter.

Roy took over the farm when he was eighteen, after Albert fell from the hayloft and broke his back. Roy and his father had never gotten along. But the bee garden was one place that seemed to humble the old man. It was a truth among beekeepers that an angry man had best not work the bees. Like most creatures, they could sense the mood in a fellow. And anger only riled them and made the work that much harder. Maybe that's why his father seemed to get a hold of himself around the bees. Whatever it was, though Albert generally begrudged his words, he did his best to pass on the secrets of working the bees to Roy during the hours they spent together in the garden. All that was long years gone. But every time Roy opened a hive to search for a queen, he remembered the old man's instructions. A queen was quick to hide herself from prying eyes. She was, after all, the life of the hive. If you knew what to look for, you could spot her special movement amid the scores of workers. Of course, you had to know where to look. You needed a sense of where she'd be at a given time in the cycle of the colony. You had to set aside your own concerns and try to feel how the bees were dealing with the day. Warm and sunny weather with the nectar flowing found them content and you could go about your business. Thunder in the air would set them on edge. A frosty morning would make them slow but defensive.

Now it was the end of another spring. This year would be Roy's last to work the bees. He planned to give them to a city-folk neighbor who'd expressed an interest. Roy and Charlene's son Tim had found them a place near him in one of those retirement communities in Florida. They'd be heading down there by the fall.

Roy made his way up the hill to the bee garden. There were still patches of snow on the ground and a chill in the air. Spring came late to that part of the county, and on Roy's hill the flowers were sometimes two or three weeks behind those in the valley just below. Roy stood and looked back down the hill to the farmhouse and the barn. The view had hardly changed from the days when he'd climb to the bee garden with his father. Seventy years and more, they'd kept

THE WAY TO BEE

bees on that hill. *Maybe it was time for them all to take a rest. A ring of electric fence wire surrounded the bee garden, there to give the bears pause before they set about toppling the hives in search of honey. Roy switched off the fence and stepped through.*

His father had taught him how to listen to the bees. You could tell their mood from the sound they made. A scratchy rattle on damp and dreary days when they were irritable and anxious from being cooped up. A baritone hum on hazy summer days when the nectar was in full flow. A confused buzz when they found themselves queenless.

Roy still remembered the first time he had helped his father in the bee garden. It was a June evening. They had finished mowing and raking the hay. They had gone up to the hives to add another layer of boxes to make room for the nectar the bees were bringing in. The bees were making honey and the air was full of a smell like warm caramel. The old man never wore gloves or a veil to work the bees, only the stained fedora he farmed in. But he put a beekeeper's outfit on Roy before they stepped in the garden.

"Aren't you worried they'll sting you?" Roy asked.

"Bees die when they sting," his father said. "They want to live as much as the rest of us, so they don't do it without good reason. You get to know their reasons, they can go about their business and you can go about yours."

In those last years, there had been mostly gruff orders or stony silence between Roy and his father. As they walked up the hill on that evening, they could see a cloud of bees rising over the garden. It was a swarm, setting out with their queen to find a new home.

The air was full of darting bees. Roy and his father stood right in their midst as 30,000 bees swirled around them. The air hummed with a sound like a strange choir. And then there was another sound. One so seldom heard that Roy had to turn and look. There was the old man, shaking his head and laughing. Laughing, Roy thought, at the pure life of it all.

And there on that same hill, thinking back, Roy laughed, too.

THE ESSENCE OF SUMMER

◆

Foraging bees fly from the hive each morning during the summer. They are searching for flowers in bloom. They fly as far as 3 miles (5 km) from the hive in their search. When they find a promising source, they sample it. They collect the nectar by sipping it through their strawlike proboscis. They then carry the nectar back to the hive in their crop or honey stomach, which is really an expandable portion of their esophagus.

WHEN THEY RETURN to the hive, the foragers deliver the nectar to other bees. The nectar itself is permeated with the scent of the flower. The bees also communicate the type of flower they have found by tactilely passing along its scent. The "hair" on the upper part of a bee's body is capable of carrying scent for a long period.

The Dance of the Bees

Bees communicate the location of the source of the nectar to their sisters through a dance. The dance of the bee communicates both the distance and direction of the nectar source. The efficiency of this communication can be observed by setting out a tablespoon of honey on a warm summer afternoon. It may take a while for the first few bees to find it. But once they do it is a matter of just fifteen minutes or so before there will be a crowd of bees jostling each other for a taste.

Solving the Mystery

The nature of the dance was deciphered by Karl von Frisch, who closely observed the phenomenon I have just described. He wondered how the discovery of nectar by a few bees could suddenly result in several hundred showing up at the same place.

> In a cowslip's bell
> Where the bee sucks, there suck I—
> In a cowslip's bell I lie.
>
> FROM *THE TEMPEST*, ACT 5, SCENE 1,
> WILLIAM SHAKESPEARE

He began by building a glass-sided observation hive. He went to the trouble of marking hundreds of bees with colored drops of pigment so that he could observe each individual bee's actions. Professor von Frisch marked the bees as they fed from a rich source of food that he set up at a feeding station some distance from the hive. He then watched through the glass of the observation hive to see what occurred.

The Round Dance

A bee would return to the hive and deliver her nectar to the other bees. Those bees would deposit the nectar in the honeycomb cells. After unloading her nectar, the foraging bee would begin what von Frisch called a round dance. She would circle once to the right and once to the left, repeating the circles with vigor. This would continue for thirty seconds or more in the same spot. Then the bee would move on to another group of bees elsewhere in the hive and resume her dance.

The bees that observed the dance would intently follow the dancer, becoming increasingly excited. At a certain point, some of the bees would fly off to the nectar source and return to perform a dance of their own. At this rate the number of bees visiting the nectar source would rapidly increase.

Interpreting the Dance

Von Frisch's observations confirmed that the bees' dance was a form of communication. Further study then enabled him to decipher their language. He learned that the round dance is performed to indicate the location of rich nectar sources near the hive. A tail-wagging dance indicates the distance and direction of more distant sources.

The Tail-Wagging Dance

In this dance a bee runs a short distance in a straight line across the honeycomb while wagging its abdomen rapidly from side to side. The bee then turns full circle to the left and runs once more. Then it turns full circle to the right and runs, repeating the pattern over and over. Von Frisch determined that the bee communicates distance by the number of turns it makes in a given time. He found that the direction of the bee's run communicated the direction to the nectar source, using the angle of the sun in the sky as a guide. He learned this by observing that the direction of the bee's run in the dance would change during the course of the day, even though

It seems as though a happy dispensation from my
scientific guiding star allowed me to discover this error
myself. But let younger investigators be warned by this
example, as they strive impatiently to publish their results
after long years of frustration. Let them test their findings
doubly and trebly before they regard any interpretation
as certain. For nature reaches her goal by another path,
where man cannot see his way.

FROM *BEES: THEIR VISION, CHEMICAL SENSES, AND LANGUAGE*, KARL VON
FRISCH, 1950: ON REALIZING A MISTAKEN ASSUMPTION IN HIS INITIAL
INTERPRETATION OF THE LANGUAGE OF THE BEES

neither the nectar source nor the hive had moved. He realized
that the changes in the dance corresponded with the height of
the sun in the sky from morning to evening.

Pass It On ...

Von Frisch also realized that the bee calculates the distance
it has traveled by the amount of energy it consumes during its
flight. Hence a shorter flight into a headwind would equate to
a longer flight in windless conditions because the energy con-
sumed would be the same. The dancing bee is basically telling
the other bees, fly in this direction until you've used up this
much fuel, then look for a flower that smells like this. Now
bring back nectar and tell the others.

Von Frisch observed that the flowers that provide the most abundant and accessible source of nectar and lasting fragrance win out in their competition for the bees' attentions. The bees will continue to pass the message along to each other until they have harvested all the nectar from a particular source. Then they will move on to another source and another dance.

MEDITATION POSTURE

◆

If you've managed to cultivate the meditation prerequisites to some degree, then it's time to take a seat. Traditional instructions describe seven or eight aspects of a meditative posture. The physical purpose is to assume a posture that enables you to remain both attentive and relaxed. Successful meditation is a matter of finding the right balance. Too relaxed and the mind is overcome by sinking. Too tense and the mind is overtaken by excitability and wandering. The purpose of developing concentrated meditation is to be able to focus and remain fully aware of the object of focus.

THIS KIND OF MEDITATION is usually practiced in a seated posture. Standing and walking are also possible, but lying down generally leads to sleep.

> *Meditation is a matter of finding the right balance.*

Legs The lotus posture, seated on the floor or on a cushion with your legs crossed and feet upturned, is advised. But if that's not possible, crossed legs in a half-lotus position is fine. If you can't sit on the floor, a chair will do instead. The knees should be slightly lower than the hips to help reduce strain and keep the back straight.

Arms The hands can be placed on the lap with the palms upward and the thumbs lightly touching. Some instructions will tell you that if you have more hatred than obsession, you should place the left hand over the right. If you have more obsession than hatred, place the right hand over the left.

Spine The back should be erect, balanced above the tailbone, not too far forward or back.

Shoulders The shoulders should be relaxed and not hunched up. Otherwise they can become a great source of tension during a prolonged sitting.

Head The head should be balanced over the spine, not tilted too far forward or backward.

Eyes The eyes should be softly focused on the tip of your nose or toward the floor about 4–5 feet (1.25–1.5 m) in front of you. They should be half-closed but still let in some light.

You can close them and feel a greater sense of concentration for a while, but this will generally lead to a sinking mind. If you keep your eyes half-open and notice you're sinking, you can slightly raise your gaze. If you find you've been taken by a wandering mind, you can acknowledge that and slightly lower your gaze to reduce wandering. The point with every aspect of the meditation posture is to find the right balance.

Mouth Your lips should be relaxed and slightly parted. Your jaw should not be clenched. Generally, you can lightly rest the tip of your tongue on the roof of your mouth to prevent the production of too much saliva, as well as to forestall thirst.

Breath The breath should be natural, relaxed, and quiet. Let it go freely and gently in and out. It is also recommended to begin by focusing on counting the breaths, from one to ten or from one to twenty-one. Our minds are in a positive, negative, or neutral state. Yet it is difficult to jump directly from negative to positive. Focusing on the breath is neutral and makes it easier to move to a positive state for meditation.

THE EFFORT IN A DROP OF HONEY

"We don't value what we don't pay for." "There are no results with-out effort." "Nothing comes from nothing." Not ironclad laws of nature perhaps. Some exceptions may exist. But they would more than likely prove the rule when subjected to analysis. Traditional teachings will suggest comparing oneself to the Buddha and then trying to account for the difference. Both are human beings with the same capabilities and potential. The difference is in the effort.

EFFORT ON ITS OWN does not guarantee results. But what determines right effort? The fourth part of Buddha's Four Noble Truths is known as the Noble Eightfold Path. The Eightfold Path consists of right view, right intention, right speech, right conduct, right livelihood, right effort, right mindfulness, right meditative concentration. Right effort summons up the energy to support positive helpful actions and let go of negative harmful actions. That requires knowing the difference between what is helpful and what is harmful.

Busy Bees

It's easy to say "busy as a bee." Are bees busy? Consider that each drop of honey represents about eighty drops of nectar. Once the watery nectar is deposited in the honeycomb cells, the bees have to evaporate nearly all of the moisture before it becomes honey. A strong working hive can contain close to 200 pounds

(90 kg) of honey. What remains is only 20 percent of the nectar the bees originally carried in, drop by drop. Even a teaspoonful represents thousands upon thousands of flights of foraging bees. One calculation has it that 1 pound (450 g) of honey represents visits to two million flowers.

> And what, monks, is right effort?
>
> (i) There is the case where a monk generates desire, endeavors, activates persistence, upholds and exerts his intent for the sake of the non-arising of evil, unskillful qualities that have not yet arisen.
>
> (ii) He generates desire, endeavors, activates persistence, upholds and exerts his intent for the sake of the abandonment of evil, unskillful qualities that have arisen.
>
> (iii) He generates desire, endeavors, activates persistence, upholds and exerts his intent for the sake of the arising of skillful qualities that have not yet arisen.
>
> (iv) He generates desire, endeavors, activates persistence, upholds and exerts his intent for the maintenance, non-confusion, increase, plenitude, development, and culmination of skillful qualities that have arisen.
>
> This, monks, is called right effort.
>
> FROM "SUTRA 117: THE EXPOSITION OF THE
> NOBLE EIGHTFOLD PATH," BUDDHA

> Like trains of cars on tracks of plush
>
> I hear the level bee:
>
> A jar across the flowers goes,
>
> Their velvet masonry
>
> Withstands until the sweet assault
>
> Their chivalry consumes,
>
> While he, victorious, tilts away
>
> To vanquish other blooms.
>
> His feet are shod with gauze,
>
> His helmet is of gold;
>
> His breast, a single onyx
>
> With chrysoprase, inlaid.
>
> His labour is a chant,
>
> His idleness a tune;
>
> Oh, for a bee's experience
>
> Of clovers and of noon!
>
> FROM "POEMS—THE BEE (XV)," EMILY DICKINSON

The foraging workers simply wear themselves out with effort over the course of the summer. You can even see the difference between newly hatched bees—with their shining golden "fur" and sparkling wings—and the worn and balding veterans, with their wings all tattered and torn. The stages of the bees' development and the roles they play during their lives intimately reflect the needs of the hive.

STAGES OF THE MEDITATOR

◆

The mind of an ordinary individual can be transformed into real-izing its greater potential through the application of practice and instruction. The aspiring meditator—determined to develop the ability to practice perfect concentrated meditation—progresses through nine stages on the way to that goal. These stages are based on nearly 3,000 years of observation of the nature and qualities of the mind. They are described in traditional teachings and depicted in illustrations of the meditator's journey.

Each stage is supported by one of six particular aspects or powers of the mind and is categorized by one of four different mental states.

Fixing the Mind

The meditator begins by attempting to maintain focus. It could be on the breath, on a visualization, or on the experience of mind itself. This attempt is usually supported by a mental power traditionally described as study or learning, where the meditator relies on the instructions of a teacher with regard to a proper object of meditation and how to engage in practice. At this stage the mind will continually waver due to either sinking or wandering. Conceptual and analytical thinking can detect the lack of focus. Thoughts seem to increase at this stage. In reality, it is insight making you more aware of the onslaught of conceptual thought. Once you recognize that the mind has left the object of meditation, the power of learning reminds you to return to the original instruction. At this stage the mental state is described as forced fixation, because the mind must be continually pulled back to focus by the power of study.

Fixation with Some Continuity

Here there is some persistence of focus. The thoughts sometimes seem more pacified. At other times they will continue to lead the mind away from the object of focus. The meditator begins to rely on the power of contemplation. This is the ability to evoke the object of meditation and remain focused on it for a time. This stage is also characterized as forced fixation.

Patchy Fixation

In this stage the object of focus persists, but the mind can become distracted by sinking and wandering. The power of mindfulness provides the meditator with immediate awareness of the encroaching distractions and enables the meditator to "patch" the meditation. Conceptual thoughts seem to exhaust themselves if followed, like a cloud evaporating in the sky. As the power of mindfulness develops, the meditator will be able to hold focus for longer periods until it lasts for the duration of the session. This stage and the next four are characterized as interrupted fixation.

Good Fixation

At this level, the meditator no longer loses the object of meditation. But the mind is still subject to the wavering of sinking and wandering. Mindfulness is still used to make the meditator aware of the signs of sinking and wandering and bring the mind back to sharp focus.

Becoming Disciplined

At this point the meditator will have developed a new mental power through the exercise of mindfulness. This is the power which is known as vigilance or meta-attention. All of us have the power of mindfulness. Meta-attention, however, is something that can be developed, but is otherwise not immediately available to the meditator. It is a part of the mind that stands guard during the meditation, apart from and not interfering with the principal focus of the mind.

Meta-attention helps buoy the mind during this stage of development, when the mind can become overly knotted and subject to subtle sinking or dullness. In this case, the mind is still focused but has lost its clarity.

Becoming Peaceful

Here the power to focus is strong but the mind is now subject to subtle excitement. The power of meta-attention serves to recognize subtle excitement and balance the mind without interrupting concentration on the object of meditation.

Becoming Very Pacified

By now the obstacles of subtle sinking and excitement are no longer a threat since they cannot interrupt for long. But effort is still required to stop them. The power of this and the following stage is known as perseverance or joyful effort, as you wish to abandon sinking and wandering because you recognize the damage they can cause if allowed to develop.

Becoming Focused

After a slight initial effort the entire meditation session is no longer interrupted by sinking or wandering. The mental state of this stage is characterized as uninterrupted fixations with effort, since the meditator still needs to make an effort to begin and end the session.

Fixed Absorption

Finally, concentrated meditation or calm abiding is effortless. Effortlessness is achieved through the power of familiarity. This stage must be practiced again and again to realize the exceptional joy that comes from the physical and mental suppleness that is the result.

FOLLOWING THE BLOOM

◆

All summer long the flowers have been blooming in their arranged progression. The bees have found fields of clover and alfalfa and raided their blossoms before my neighbors come to mow the hay. I see them everywhere I go. I find them sipping water from hoof- prints. They're on the tiny wildflowers in the pasture, visiting whatever blooms appeal to their relentless sense of purpose. Is it possible to see in this not just one summer but all summers? Not just one season but all seasons? Not to deny their uniqueness, but to realize that it is all in the nature of change? Nothing to grasp. Nothing to cling to. Nothing to fear. Everything to appreciate.

THE DAYS HAVE BEEN GROWING steadily shorter since the solstice in June. The evenings are chillier once the sun has set. The season is winding down.

The last flowers before the frost will be the goldenrod and the aster. The bees are determined to gather every last drop before the first frost of fall. Soon they will make their final foraging flights of the year. Soon I will weigh the hive boxes to see what the summer has brought. But right now it's time to sit beside the hives in the warmth of a September after- noon and listen to the buzzing of the bees.

The bees are determined to gather every last drop before the first frost of fall.

My confidence in venturing into science lies in my
basic belief that as in science so in Buddhism,
understanding the nature of reality is pursued by means
of critical investigation: if scientific analysis
were conclusively to demonstrate certain claims in
Buddhism to be false, then we must accept the findings
of science and abandon those claims.

FROM *THE UNIVERSE IN A SINGLE ATOM:*
THE CONVERGENCE OF SCIENCE AND SPIRITUALITY,
H.H. THE FOURTEENTH DALAI LAMA, 2005

Listening to the Bees

I think we need to exercise patient observation and clear-eyed
analysis. We must allow ourselves to test assumptions "doubly
and trebly" before drawing a conclusion. I believe both Buddha
and von Frisch would be in accord with these instructions.
In this regard, the meditator and the scientist are not so far
apart. The wealth of what each discovered in his own search—
one for the nature of reality and the other for nature—
is undeniable. Both were guided by a desire to find the truth.

CHAPTER THREE

FALL

*The autumnal equinox has arrived. From now
until the solstice, the nights will outlast the days.
Andromeda, Cassiopeia, and the hero Perseus take their
places in the northern sky. To the south, Capricorn
pursues Sagittarius across the horizon. The mornings
dawn crisper. The days turn brisker. In the orchard the
apples that began as blossoms now weigh down the
boughs. The coming harvest also marks the end of
increase. Each of fall's flowers is a reminder that
every drop of nectar is precious. The bees know
this and prepare themselves. Time will judge their
efforts soon enough.*

THE CODE OF JUSTINIAN

I wasn't always the happy camper I claim to be these days. After my brush with fame and fortune, things unravelled a bit. I found myself living back in New York in my old apartment and working as a truck driver to pay the rent.

On one of my trips out to New Jersey I passed a sign that said "Local Honey." It was in front of an old farmhouse left over from the days when the whole area was farm fields instead of tract houses. I pulled into the driveway. I could see a bunch of beehives behind the house. I got out and knocked on the front door. There was no answer. I walked around the back. There were at least thirty hives set up in rows on a small rise behind the house.

There was a long, low building on one edge of the property. It looked like it used to be a chicken coop. The building sat on blocks above the ground. A series of trapdoors were cut into one wall at floor level. Bees were flying in and out. I figured the owner had set up some hives inside the building to protect them from the weather.

I went to the door of the shed and knocked. I heard somebody call out, but I couldn't make out what they said. I waited by the door. There was a stand of goldenrod nearby. It was fall and I could hear the bees buzzing in the blossoms.

There was a pond at the end of the property at the foot of the rise. A pair of Canada geese was swimming in the pond. I'd heard somewhere that they mated for life. If one was injured and couldn't fly when winter was coming, the other would just stay there, even if it meant freezing. That was my idea of real romance.

The door to the shed opened. A white-haired old guy in a beekeeping veil stood there. He was dressed in khaki work clothes. Like most veteran beekeepers, he wasn't wearing gloves.

"I saw your sign for honey."

"My wife usually makes the sales, but she's out at a singing rehearsal. They're putting on a show at the town hall the end of the month and she don't like to miss her singing. You interested in one-pound or five-pound jars? Or I've got comb honey, but I don't care to make much of it. Uses too much wax. Bees work hard enough without cutting out their comb. But some customers will only go for the comb."

"Actually, I was wondering if I could have a look at your setup. I keep a few hives myself."

"Hobby hives?"

"Something like that."

"Had any trouble with the law?"

"Not yet."

"With all these developments moving in, people are afraid they'll get stung. They raise all kinds of hell with the township. Our place is still zoned agricultural or they'd have shut us down for sure already. As it is, the wife's always worried some kid will get stung and we'll get sued."

"You ought to tell her about the Code of Justinian. He was the guy who put together the laws for the Roman Empire. He said if somebody's cattle strayed on to your place and stomped on your garden, you had to return the cows, but you could collect damages because you could identify where the cattle came from. I guess they had brands or something back then. But if a swarm of bees landed on one of your trees, you could keep them even if the beekeeper next door was jumping up and down. Because there's no way to prove they're his bees. I figure it's the same with a sting. As long as the kid's not trespassing on your property, how can they prove the bee came from your hives?"

"I told her as much myself. But she figures they'd sue us anyway and we'd have to pay the lawyers even if they couldn't prove it."

"She's probably right there," I admitted.

His name was Herman West. His family had farmed around there for a couple hundred years. But the farms were all gone now, sold off field by field as the kids turned their backs on farming and moved away. Herman told me he sold off his remaining cornfields when he retired, after he broke his hip on the ice a few winters back. The developers filled them with houses practically overnight. Herman kept the orchard and a few acres so he could still keep bees and get the agricultural exemption on the murderous property taxes.

After a little more chit-chat, he showed me his hives. It was a warm day and the bees were going full tilt, buzzing in and out of the hives. He had about sixty colonies on the property altogether. He said he had another thirty at a cousin's farm at the other end of the county. He got about 75 pounds (34 kg) out of a hive in a good year. That worked out to over three tons of honey. He still managed it himself, but it was getting to be a lot of lifting, with his hip and all. He had invented rigs to help with moving the boxes and harvesting the honey. But it was still lots of work, and his wife was always after him to cut back on the operation. As it was, he had to sell most of the honey in bulk as there were too many jars to move locally. He could get a much better price if he sold it by the jar, but he wasn't up for that much traveling.

I told him maybe I could help him out. I was making the rounds of the grocery stores anyway. If he wanted to cut me in on the sales, I could help him with some of the beekeeping and shop the honey at the markets I visited. He could see from our tour of the hives that I didn't mind standing in the midst of a cloud of bees. He did a little calculating and figured the difference was worth it. So we made a deal.

Thereafter, I'd stop by most weekends and help him with the chores and then I'd set off with some cases of honey for the markets in the area. I tried to cover as much territory as possible.

I became known as the beeman. I'd make the rounds from shop to shop. I'd sell them a case or two of honey, then head off to the next shop. Sometimes they'd sell out and call me for more. I didn't like going back over the same ground unless they called because I was trying to move three tons of honey.

I liked the smaller markets. The ones owned by someone local instead of those giant impersonal chains. In general, they were a lot more easy-going and they could buy the honey right away without going through a lot of red tape.

I was looking for something on those drives. I had some idea that maybe I'd run into the answer; as if finding my way had something to do with the amount of mileage I did.

Perhaps I never found my way, but it did seem like a good idea at the time.

GOLDEN HARVEST

◆

The time has come. The first hard frost has killed the last of the season's flowers. As the supply of nectar dwindles then ceases, the bees become defensive. What they have gathered must sustain them until next May, when the nectar flows again and they've had enough time to make some honey. The queen has slowed her egg laying in anticipation of the coming winter. The population will begin to diminish as older bees die off, not to be replaced until spring.

> Bees work for man, and yet they never bruise
> Their Master's flower, but leave it having done,
> As fair as ever and as fit to use;
> So both the flower doth stay and honey run.
>
> FROM "THE TEMPLE—THE CHURCH—PROVIDENCE,"
> GEORGE HERBERT, 1633

IT'S TIME TO TAKE STOCK. All summer long I have been adding the boxes called supers to the top of the hives. Each box contains nine or ten frames of empty honeycomb. Each super holds about 25 lbs. (11 kg) of honey when the bees have filled the frames. As the bees fill each super, I add another.

As the season draws to a close, it takes them longer to fill each box. The colder weather and shorter days mean a later start each morning for the foragers and more time is required to evaporate the moisture from the nectar.

Weighing up the Success

Now that the frost has come I pull off the supers and stack them to the side for a moment so that I can gauge the weight of the two lower hive bodies. The bees need about 100 pounds (45 kg) of honey to make it through a cold winter in New York.

I lift the boxes, hoping it will be a strain. If they feel full and heavy, I know I can take the supers for harvest. If they are disappointingly light, I have to leave some of the harvest for them. Sometimes I have to ensure that the weaker hives have enough by giving them a super from a stronger one.

Right now the supers are filled with bees. The supers need to be free of bees when it comes time to harvest. If not, the experience could be unpleasant for both the beekeeper and the bees.

Exclusion Zone

I use a device called an excluder. It's a slim box with a series of small plastic cones that function something like a lobster trap. The wide section of the cone opens into the super. The narrow section exits into the hive body underneath. I place the excluder on the hive body and then set the bee-filled supers on top.

On a cold evening, the bees in the super will retreat down to the warmth of the lower hive. The shape of the cone allows them easy travel, but only in one direction. The narrow opening back to the super is just small enough to prevent them from returning. After one or two cold October nights, the bees are all below and the supers are ready to be removed.

ANALYTICAL MEDITATION

In the first part of this book I explained the benefits and stages of developing concentrated meditation or calm abiding. If we had all the time in the world, we could proceed through those stages at our own pace and reach the ninth level of effortless fixed absorption, and then proceed to perfect analytical or special insight meditation. But of course, in reality, time is short. With this in mind, it might be a good idea to develop the benefits of both concentrated and analytical meditation in tandem. Focusing on the nature of impermanence can be a good start.

F IRSTLY, MEDITATING ON IMPERMANENCE can have several benefits, as I see it. It can help me appreciate the opportunities that I have. It can make me more aware of the urgency needed to take advantage of those opportunities, which could become unavailable at any moment.

Secondly, understanding impermanence can lessen my attachment to objects of obsession by realizing they cannot provide a permanent source of happiness.

Possessions and Attachment

Attachment can drive us to distraction. We can't let go and we can't get enough. We stake our happiness on our obsessions and we derive our misery from them. Attachment robs of us the ability to appreciate the good things we have. It is difficult to appreciate something if you live in fear of losing it or are driven to acquire ever more. This just produces constant turbulence in our minds.

The things we try to hold are all impermanent. My house in the country is slowly but surely returning to the earth. The roof weathers and leaks in the winter. I can repair it but then the boiler breaks down. I can fix that, too, but the trees in the front yard are aging. I can plant new trees, but will I be around long enough to see them grow?

People and Attachment

How much more impossible is our attachment to others? In my heart I behave as if people are there to make me happy. But to make me happy they have to conform to my idea of how they should behave. If I judge them by my standards instead of accepting them for themselves, I will be constantly disappointed. If they turn in their affections, I am either outraged or devastated.

Attachment steals our contentment. The Buddha taught that our chief troubles come from not being content with what we have. The antidote is recognizing the impermanence of what we desire. By relaxing our fearful grasping we can be free to appreciate what we have before us.

The Buddha taught that our chief troubles come from not being content with what we have.

KNOWING THE MAIN THING

My friend Roy Kaiser the dairy farmer had retired. He was spending his time tending the garden and playing with his grandchildren when they came to visit. He had trouble with his shoulders and it had reached the point where he could hardly grip a coffee cup and raise it to his chin. He was just worn out from a lifetime of hard work. His wife Charlene fussed over him. She tried to keep her worries to herself, but Charlene was never much good at hiding her feelings. Roy didn't complain and he let Charlene go on with her fussing since he figured she felt better that way. They were married for forty-seven years; he told me that trying to do what he could to make Charlene happy was the main thing.

THE PROPERTIES OF HONEY

◆

Honey's low water content and relatively high acidic level make it unfavorable for bacteria and other microorganisms. The bees remove most of the moisture from nectar to make honey. This makes honey hygroscopic, meaning that it readily absorbs moisture. Bacteria cannot live in honey because it draws the water from their cells. Yeast cannot survive in honey as long as the moisture content is below 17 percent. This quality also means that honey does not spoil as long as it is not exposed to moisture.

HONEY IS A SUPERSATURATED SOLUTION. It contains more dissolved sugars than can normally remain in solution. This means that honey can crystallize, depending on its sugar composition, moisture content, and temperature. Some honey will crystallize while still in the comb, but other honey will not crystallize for years. I have some jars of exceptionally dark and flavorful honey that have not crystallized in fifteen years.

> The pedigree of honey
> Does not concern the bee;
> A clover, any time, to him
> Is aristocracy.
>
> FROM "POEMS (V)," EMILY DICKINSON

Bacteria cannot live in honey because it draws the water from their cells.

Why Eat Honey?

Honey contains antioxidants. Antioxidants can protect the body against biologically destructive chemicals known as free radicals, which have been linked to skin aging as well as many diseases. Studies have shown that darker-colored honeys such as buckwheat seem to contain more antioxidants than lighter-colored varieties.

The antioxidants in honey can help the skin cells recover from damage caused by exposure to the sun, air pollution, or other causes. That's not all. Not only can these antioxidants help to eliminate free radicals in the body, but they are also among the nutrients that promote the growth of new tissue.

The Rewards of Effort

It's time to extract the honey. Time to measure at last the reward of the bees' prodigious efforts and our far more modest labor. It is time to taste the essence of the summer.

I bring the supers into the shop the night before. I make a fire in the wood stove and try to get the temperature in the room up to about 90°F (32°C). It will warm the supers as they sit overnight and help the honey flow more easily when it's time to extract it in the morning.

The next day the shop is full of the aroma of honey. I load in more wood until the stove is roaring. The bees have capped the honey in the comb with a layer of wax. The extraction begins by removing a frame of honeycomb from the super to cut off the capping. I use a heated knife to slice off the capping. The wax cappings then go into a melter. I use the wax in the winter to make candles.

"We're Going to be Rich"

I put the frames into a hand-cranked extractor. The extractor spins the frames and the centrifugal force throws the honey from the cells in the comb. Friends and neighbors come by to watch or lend a hand.

Slowly the extractor begins to fill. When the honey is just below the level of the spinning frames, it's time to empty the first bucketful. I place a 5-gallon (19-liter) bucket under the extractor valve. This is the moment my six-year-old son is waiting for. He opens the valve and the honey streams into the bucket.

"We're going to be rich," he says.

"We already are," I reply.

Harvest Home

It takes most of the day to uncap and extract all the frames. When I'm done extracting, I put the frames back in the supers and stack them outside. If it's a warm afternoon, the bees will show up in a matter of minutes.

In a day or so they will have cleaned out every last bit of honey. The comb and the frames will seem as if they never held a drop. I let the honey sit in a holding tank overnight so that all the air bubbles can rise to the surface.

The next morning we're ready to bottle the honey. I fill each jar to the top and label it, then stack them twenty-four to a box. The honey is still warm.

Sharing the Riches
A few hours more spent cleaning up. And that's it. The essence of a summer's flowers, in jars.

Each year's harvest is different. Hundreds of pounds one year. Thirty the next. The honey can be light and golden or as dark as molasses. It depends on the season's weather and which flowers have won out in their courtship of the bees.

I share it with family and friends.

I don't suppose it will always be like this. How could it? Even though I've repeated this harvest for many years running, the years themselves are a sign the time is running out. Increase. Decline. Change is in the nature of things. But for now it's enough to set a jar of this year's honey on the table and appreciate another season.

> A spoonful of honey will catch
> more flies than a gallon of vinegar.
>
> FROM *POOR RICHARD'S ALMANACK*,
> BENJAMIN FRANKLIN, 1744

URSA TEACHES EQUANIMITY

◆

The bears have ravaged the hives in the fall for several years. Some sceptered isles are free from this onslaught. But not the mountains where I keep my bees. I seem to take each sling and arrow of this outrageous fortune personally. I may try to mask this by telling myself it's the bees in my charge that I feel for. The truth is that I'd like to blame and banish the bear. I've already convicted and sentenced him as a trespasser, thief, and violent offender. How's that for an anthropomorphic, egocentric point of view?

I'M ALWAYS LOOKING for someone to blame for launching outrageous fortune's slings and arrows at me. This kind of finger-pointing can seem like a full-time occupation. Many an interior monologue has made this the subject for defense and counter-attack. How many invisible armies of thought have I mustered in defense of my imagined realm?

> Seeing that the chronic disease of self-cherishing
> Is the cause of my unwanted suffering,
> Inspire me to put the blame where blame is due
> And vanquish the great demon of clinging to self.
>
> FROM "OFFERING TO THE SPIRITUAL GUIDE,"
> LOBSANG CHÖKYI GYALTSÄN, 1570–1662

Nothing to Defend

I watch the female workers dragging out the resisting drones after the frost has killed the flowers. I project a feminist victory over all the deadbeat dads. I impose a work ethic and entitlement over a process of reproduction and resource conservation that evolved long before any arguments started about rights of wealth distribution.

The bees may teach order to a peopled kingdom, but they also demonstrate to me a lot about my predispositions and possessiveness. This is the kind of habitual thinking and closed-mindedness that real researchers struggle to overcome.

In the end I strongly suspect there is not all that much to defend. At least not as I habitually regard it. I have long favored a quote attributed to both Benjamin Disraeli and the industrialist Henry Ford II: "Don't complain. Don't explain."

A Meditator's Trick

My teacher Gelek Rimpoche suffers from diabetes. He has to inject himself with insulin every day. I once asked him if it hurt. He said of course it does. But he also said that since he has to suffer from diabetes anyway, he might as well not suffer uselessly. So when he's injecting himself he prays that he may suffer diabetes for all the diabetics in the world, so that they may be free from the suffering of diabetes from now on. Rimpoche called that a meditator's trick. I call that chasing the thief on his own horse.

The Zen master Hakuin was praised by his
neighbors as one living a pure life.
A beautiful Japanese girl whose parents owned a food
store lived near him. Suddenly, without any warning,
her parents discovered she was with child.
This made her parents angry. She would not confess
who the man was, but after much harassment at
last named Hakuin.
In great anger the parents went to the master.
"Is that so?" was all he would say.
After the child was born, it was brought to Hakuin. By
this time, he had lost his reputation, which did not trouble
him, but he took very good care of the child. He obtained
milk from his neighbors and everything else he needed.
A year later the girl-mother could stand it no longer.
She told her parents the truth. The real father of the child
was a young man who worked in the fish market.
The mother and father of the girl at once went to
Hakuin to ask forgiveness, to apologize at length, and
to get the child back.
Hakuin was willing. In yielding the child all he said was:
"Is that so?"

FROM *ZEN FLESH, ZEN BONES*, PAUL REPS, 1961

FINDING THE WORDS

I went to a shrink once because I had nothing better to do with my money. I was dating someone at the time and she made it a condition. It was one of those let's-discuss-our-problems-to-the-exclusion-of-all-other-topics relationships. I was confused enough at the time that the proposition made sense. The problem was I had nothing to say.

At the end of my first meeting with the shrink he left me with the words, "You look like a man who's carrying a heavy burden." I had no idea what he was talking about, being such a happy outgoing person and all. But little by little I began to feel the weight of it. Still, I stayed speechless. It went that way with him, week after week, struggling to find the words. Sitting in a shrink's office at a loss for words can really get a guy down. I know it's part of the process, but it didn't leave me feeling any better.

So I left the shrink and went to see this crazy Chinese guy who was supposed to be able to walk through walls. He was trained in the mysterious magic arts; now that's my idea of fun. However, I'm not saying there isn't more than meets the eye in the magic arts; I'm just saying that I never saw him use anything besides the door.

Now the Chinese guy might not have been able to walk through walls at will, but he could see right through me. Not that it's such a hard job, really. How much do we really hide when you get right down to it? It's more a question of how much we want to see. I told this guy about my being tongue-tied about pretty much everything. He told me two things. When you've been shot with an arrow and you can't pull it out, sometimes you just have to push it through. The other was that since I had such a hard time saying what was in my heart, I'd better start practicing. Someday I might need to find the words.

The Wait Begins

With the last of the nectar stored and the nights growing frosty, the bees settle in for the coming winter. Their stored honey and their fertile queen must be preserved and protected if the hive is to survive until spring.

As the first snow flurries fall in November, the bees cluster together and wait.

CHAPTER FOUR

WINTER

There is a Tibetan saying: "If you want to know your past, look at your present conditions. If you want to know your future, look at your present actions." Intention, preparation, effort. These are the factors that govern the results of our actions. Whether they are done poorly or with determination—with a positive or negative attitude—affects how we will fare. It's far better not to wait until the snow flies to realize winter is coming.

TEACHING THE ART OF ORDER
TO A PEOPLED KINGDOM

I was still lingering in my down-and-out days. By that point, I had a couple of hives on the roof of the tenement in New York City. My landlord Patsy was OK with getting a share of honey as rooftop rent.

It turned out his old man kept bees on the roof when he was a kid. Patsy had grown sentimental in his old age. He was too old to enjoy the five-floor climb to the roof. But on sunny days, he'd sit in a chair that one of the guys from messenger service on the first floor would drag to the sidewalk. He'd reminisce about the old days in the neighborhood when most of the buildings had bees on the roof and all had chickens and goats out the back. Everybody had a garden where they'd grow grapes and make grappa in the basement. He talked about it as it used to be—one big, bucolic farmland instead of the half-step up from a slum it became before the real-estate agents discovered it.

I kept up my old routine. In the mornings before work, I'd go to the roof and check the bees. They were adjusting to city life better than I was, busy bringing in pollen and nectar from who-knows-where— probably from the new penthouse gardens springing up all over the neighborhood. I liked to watch bees work. They seemed so full of purpose. A steady stream of bees would land at the entrance and then make their way into the hive with their harvest. Others would circle up into the air on their way to find more blossoms. From dawn to dusk they never seemed to let up.

During the day, I'd work my truck-driving job. In between runs, I'd hang out on the corner with the landlord or whoever else was inclined to while away their time in idle conversation.

At the end of the day, I'd walk along the river. I might get a sandwich and eat it on a bench overlooking the water. I rode the ferries. I'd bring a book and read on the boat. I would just sit on the ferry and ride back and forth until I was tired enough to go to sleep.

Friday evenings I'd go up to the Metropolitan Museum of Art, when it was open late and the galleries were empty of tourists. I was not the first person who had a few setbacks in midstream. That's the reason I haunted the Met. All those old painters had spent their lives searching for something, some kind of answer to make sense of the world around them. I thought maybe they'd left a clue.

Two cops knocked on my door one Sunday after I'd gotten back to the city after making the honey rounds for Herman. There's something about the way a cop knocks that you know right away who it is.

I opened the door.

"You the guy with the bees on the roof?"

I was feeling fed up and it must have shown.

"Why? You looking to buy some honey?"

It turned out that according to Article 161.01 of the city health code, no person shall sell or give to another person, possess, harbor, or keep wild animals identified in subsection B as any animal naturally inclined to do harm and capable of inflicting harm upon human beings, including dingos, pole cats, jaguarundis, grizzly bears, kinkajous, gorillas, coral snakes, and something called the Bornean earless monitor.

The cop had his statute book with him and I made him read the whole thing until it got to hornets, wasps, and bees. I told him no one in his right mind would or could keep hornets or wasps. But bees were quite a different matter.

The code said wild animals. Bees weren't wild. They were as domestic as cows and house cats. That's why you could buy honey in the supermarket. Because they were agricultural. Nobody would eat what wasps brought home, which was dead bugs.

"That's fine," one of the cops said. "But the code says bees. That means they've got to go."

It turned out some benighted soul in the neighborhood had called in and made a complaint. No doubt it was one of the recently arrived loft owners unwilling to share the scent of flowering locusts with those who could actually make some use of it. I suppose it was for the likes of these that such laws were made.

"I'm not going to play games with you guys," I said. "You're just doing your job, right? But I'm paying to keep these bees here. You want to give me a summons or haul me off or shoot me, fine. It's all the same to me. But I'm not moving them."

I was calm when I said it. But I resented the wave of designer kitchens and bottled water and to-die-for shoes that was washing over the neighborhood, reminding me of the high life I had once led. I wasn't going to let go without a fight.

The cops wrote up a summons and left it at that. It was clear they didn't want to mess with the bees themselves, despite my claims of their domestic pedigree. And it probably wasn't worth the paperwork to haul me in or shoot me. They also issued a summons to Patsy, my landlord. That turned out to be a mistake.

"Who do they think they are, writing me up?" he asked me the next afternoon. "My old man kept bees here. Right on this same roof. Same as calling my old man a criminal. It's my building. If I want to level it to grow tomatoes, no city-hall son of a ballerina is going to tell me how to water them." Or words to that effect.

I told him my theory about the Code of Justinian.

"Yeah, I remember hearing something about him from the nuns," Patsy said. "He's the one who saw some cross in the sky and turned all the Turks into Catholics."

That was Constantine, but I didn't feel it was going to be worth my while making the distinction.

"I already got my lawyers on this," he said. "You want, you go down and talk to them. I'll tell them to take care of you along with this summons. I find out which one of these flower arrangers made the call, I'm gonna make them a non-refundable offer to move to the suburbs."

I told him I'd prefer to handle my end of it myself, as long as he had no objection to me keeping the bees on the roof until it was resolved one way or the other.

He said it was my call about the lawyers. As far as the bees were concerned, I was paying good rent for them and it was up to him who stayed and who went in his buildings.

It turned out his lawyers were the same ones oil companies and the rich Arab nations hired when they had a little problem with parking tickets or capsized crude carriers. I guessed New York City real estate required a little more legal finesse than your average enterprise. Just the fact that they took on the case meant it could easily drag into the next century. For the moment the beehives remained where they were.

One of the complainants must have called the news to rant about how Patsy was harboring dangerous animals with total disregard for the safety of the neighborhood. Patsy told the reporters he was considering raising buffaloes in the lot where he parked his trucks. If a buffalo was good enough for the government to put on a nickel, it should be OK with the rest of the (expletive-deleted) citizens.

Patsy had picked up on some of my rambling about bees. He told the reporters that, unlike wasps, which are carnivores that hunt and kill other insects, honeybees live off nectar and pollen and die when they sting. Because it's so costly, honeybees only sting in defense of their hive. He said that, like most of us, if you leave them alone, they'll leave you alone. But if you started messing with them, you'd better look out, because they'd do whatever it takes to protect their patch.

SEASONS COME FULL CYCLE

◆

The goal of the bees is to survive the winter. The task of the bee-keeper is to wait. I can dip beeswax candles. Make mead. Stir some honey into tea. Build new hive boxes and repair old equipment, but this idle work only highlights the fact that there's little to do but wait and see how the bees will fare the cold. Otherwise it's time spent wandering out to the bee garden looking for dead bees in the snow and tapping on the side of the hives to listen for signs of life.

> They alone know a country, and a settled home,
>
> and in summer, remembering the winter to come,
>
> undergo labor, storing their gains for all.
>
> FROM *GEORGICS, BOOK IV, BEEKEEPING (APICULTURE),*
> VIRGIL

T HE HONEY HARVEST IS OVER. It's time to ready the hives for winter. Mainly that means ensuring they have enough honey stored and protecting them from the winter wind.

It's time to put on the entrance reducers to keep the mice out. To weigh down the covers with stones to keep them from blowing off during the winter gales. To place hay bales around the sides of the hive for insulation.

Mainly, it's a time to wait.

Long Nights Ahead

In January, I place the order for the new bees. The feeling is definitely mixed. I'm ordering replacements because I am pessimistic about how many hives will make it through the winter. I'm hopeful because I think the new bees will have a chance. The price of a new package of bees has gone up steadily since I first began practicing the art. If beekeeping were strictly a business, I would have had to give up long ago, like my commercial neighbors.

Silence

Winter is a silent time in the mountains. The croaking frogs have long since buried themselves in the mud of the pond. Most of the birds have migrated south to warmer climes, leaving behind a few cawing crows and some chickadees that only accentuate the silence.

I have spent some winter retreats at the monastery hidden at the end of the mountain road not far from my house. Sometimes a silent snowfall would blanket the zendo (the place for study and meditation) with 3 feet (1 m) of snowfall, making the quiet all that more profound.

> To the mind that is still, the whole universe surrenders.
>
> LAO TZU

LISTENING TO THE SILENCE

I've never had that much to say. My good friend Roy Kaiser was in the same boat. One time I dropped by his farm and found him alone in the house. Charlene was out on an errand somewhere. Roy and I sat together in the kitchen drinking coffee. Charlene is usually the one who holds up both ends of the conversation. Her stories take you down the side roads, through the woods, and up over the hayfield before you find your way to the moral of the tale. Roy's part is generally played with a bare word or two. So when Charlene walked in the back door and found us sitting at the table, she stopped in her tracks with a look of staged astonishment.

"Well. How long have you two been sitting here?" she asked.

"About an hour," Roy answered.

"And whatever did you find to talk about?" she said.

Roy waited a moment for the dust to settle.

"We were listening to each other," he answered.

DEFENDING AGAINST PREDATORS

◆

The hives need protection in the winter from their chief predators.
I keep my hives in upstate New York. Skunks are one minor predator.
In winter the skunks will scratch on the front of the hive to alarm
the bees. You can see evidence of the skunks' scratch marks on the
front of the hive if this has occurred. Some bees will sally forth to
defend the hive, only to be eaten by the waiting skunk. I can only
begin to imagine what a mouthful of bees might feel like.

ONE WAY TO KEEP THE SKUNKS AWAY from the entrance
in the winter is to place a board of nails in front—
something like a small bed of nails—which makes it difficult
for the skunk to get close enough to scratch the hive.

Field mice also like to nest inside the hives during the
winter. The bees keep the hive warm with the heat they
generate as they metabolize honey. The mice no doubt find it
a cozy place to curl up. However, in building their nests, the
mice chew through large sections of comb. If the population
of the hive is strong enough, the bees will try to drive the
mice from the hive.

But in the winter months, when the bees are clustered
together for warmth in the upper story of the hive, the mice
may go unmolested. However, if a mouse should die within
the hive, the bees will embalm it with propolis. The bees are
meticulous housekeepers. They carry out their own dead and

drop them 3 feet (1 m) or so from the hive entrance; however, a mouse is too large to move, so the bees will embalm it in place. The usual precaution against mice is for the beekeeper to reduce the size of the entrance for the winter with a wooden or metal guard called an entrance reducer.

Bears Love Honey

The black bear is the main predator where I keep my bees. Here is where the defense grows more challenging. True to the children's tales, there is little doubt from what I've seen that bears love honey. And they will go to great lengths to get it from the hives. They also like to eat the young larval brood being raised in the honeycomb. The result of a bear meeting a hive looks something like the aftermath of a tornado.

I have tried strapping my hives together with metal brackets and bolting them to large wooden pallets. When full of honey at the end of summer, the entire assembly can weigh close to 300 pounds (136 kg). I have several times seen where a bear has lifted, dragged, and then demolished the entire hive some 60 feet (18 m) from where it was located.

I have tried the electric fencing my neighbors use to enclose their horses and cattle, only to find it torn apart. I have tried running the wire to the metal outer covers of the hive to no avail. I know of several beekeepers who have tried suspending their hives in the air with cables and winches in a desperate attempt to keep the black bear at bay.

Desperate Measures

After the most recent round of destruction, I retreated to a three-sided shed I had previously used as a run-in for horses to shelter from the weather. I fortified the open side with some strong metal cattle gates.

I covered the space between the gate rails with a crisscross of barbed wire. There was barely enough space to poke your head through. And even if you could, you'd come away with a faceful of unpleasant scratches. All seemed well as I took a well-earned rest from my labors.

The next morning I went to check my handiwork. From a distance all appeared in order. As I drew close I saw one hive inside this fortress had been tipped over and broken apart. I stood and stared, trying to puzzle what had happened.

The only animal large enough to tip over a hive but small enough to get through the barbed-wire gate was a raccoon. A raccoon was large enough to tip over the hive, but it wasn't strong enough to smash apart the wooden hive bodies.

In this case, one of the ¾-inch (20-mm) pine boards had a section torn from it, as if something had taken a huge bite from the wood. That's when I saw the clumps of wiry black fur caught in the fence barbs. Lots of fur.

Somehow a bear had squeezed in through a jagged hole no bigger than the size of your head. Not only squeezed in, but had a meal of honey and squeezed out again, leaving enough fur behind to stand as evidence in a court of law.

I've since fortified the gates with wire mesh and connected a cattle charger to the shed. The state requires that apiaries be registered with the Department of Agriculture and that the registration number be prominently posted somewhere in the apiary. I've complied with the state's requirements, but I'm going the extra distance and fashioning a sign to attach to the front of the building. It reads "Folly."

BALANCED ON THE BRINK

—◆—

Virgil makes it clear we are not the first beekeepers to suffer loss. From angry nymphs to colony collapse disorder, it has been a struggle. In working with nature, anyone who has farmed for a living can tell you who has the upper hand. The effort to wring fruit from our labors has always faced challenges.

> "Son, set aside these sad sorrows from your mind.
> This is the cause of the whole disease, because of it the Nymphs,
> with whom that poor girl danced in the deep groves,
> sent ruin to your bees. Offer the gifts of a suppliant,
> asking grace, and worship the gentle girls of the woods,
> since they'll grant forgiveness to prayer, and abate their anger."
>
> FROM *GEORGICS, BOOK IV, BEEKEEPING (APICULTURE)*, VIRGIL

Balance has been a theme of this book. But balance and stasis are not the same. Things change. Finding balance means accommodating change. That's the basis of adaptation.

Brother Adam's search for a disease-resistant bee led to hybridization. Successful hybridization leads to dominance and a reduction of diversity. Lack of diversity makes the remaining species more susceptible to a single disease. There seems to be no escaping Robert Burns and his plow.

Finding the Balance

In terms of meditation, there must be a balance between attention and relaxation, between focus and analysis.

Tibetan teachings provide an example of the balance that is required in order to skillfully use the tools of meditation. Some of the temples of the old Tibetan monasteries were dark, even during the day. To see the details of the wall paintings, one needed a candle or, more commonly, a butter lamp. But if there was a draft, the lamplight would flicker, making it difficult to see without a stable, steady illumination.

In the same way, concentrated meditation stabilizes the ability to utilize the penetrating insight of analytical meditation. It is the balance of the two that brings meditative results. The goal is to see past obscurations into the nature of reality.

Finding balance means accommodating change. That's the basis of adaptation.

What Is at Stake

There are a lot of reasons to meditate. There are a number of reasons why one would want to keep bees. Whatever the reasons, one can ask a few simple questions about the results. Am I making things better or worse? Am I causing more harm than good? Am I serving others as well as myself? Determining the answers may not be so easy. But asking the questions is essential.

The lessons taught by a teacher with a positive motivation penetrate deepest into their students' minds. I know this from my own experience. As a boy, I was very lazy. But when I was aware of the affection and concern of my tutors, their lessons would generally sink in much more successfully than if one of them was harsh or unfeeling that day.

So far as the specifics of education are concerned, that is for the experts. I will, therefore, confine myself to a few suggestions. The first is that in order to awaken young people's consciousness to the importance of basic human values, it is better not to present society's problems purely as an ethical matter or as a religious matter. It is important to emphasize that what is at stake is our continued survival.

FROM *THE ESSENTIAL DALAI LAMA: HIS IMPORTANT TEACHINGS,*
H.H. THE FOURTEENTH DALAI LAMA, 2006

THE GIFT OF FLIGHT

My back was against the wall. I was standing on Bill Klein's balcony high above East 70th Street in New York City.

It seemed an odd friendship at first. Bill was over ninety, and I was still in what I thought was the middle way of my life. Bill was a retired publisher and prominent consultant to the textile industry. He had been a man of the world, an elegant dresser, a connoisseur of oriental art, and an artist himself. He was one of the first Americans to visit the Zen monasteries in Japan after World War II.

I liked Bill the moment I met him, during an interview for a show on twentieth-century textiles. I knew nothing about textiles, but I was willing to wing it for the money. Bill could recognize and appreciate someone skilled in getting away with it. We continued to meet after the article had been put to bed.

I tried to visit Bill at least once a week. We would sit together in his apartment and talk. An automobile accident some years earlier had robbed Bill of his lifelong athletic strength, although this same strength had enabled him to recover from injuries that a younger man might not have survived. Although Bill was now frail, he maintained a keen interest in current events. On days when his eyesight troubled him, he would ask me to read him the newspaper.

At first I kept my own affairs to myself. Bill was a polite man, who never asked me much about my past, despite us having been on first-name terms for many years. He was more interested in the things I could do. Practical things like keeping bees or fixing taps. He didn't mind telling me about his own early days in business, his love of painting, even his affairs de cœur. I enjoyed listening and I didn't think I needed to waste his time on my little ups and downs.

I've never been much of a swimmer. That's why I like to sail. I prefer being on top of the water, not in it. Yet back then, I was feeling like a guy who had slipped overboard in the middle of the wide Sargasso Sea. I finally mentioned how I felt. As soon as I did, I had a sense of how shallow my problems really were. Here I was, frantically treading water. I never thought to see if it was shallow enough to stand.

"What do you hope to gain?" Bill asked.

"Peace, understanding. Something like that."

"How long do you generally sleep?" Bill asked quietly.

"Four to five hours. I get up at five-thirty to give myself time to work."

"That's good, as long as it doesn't interfere with your health," Bill said.

"There's just too much to do otherwise and it's hard to find the time," I said.

Bill rubbed his hand over his balding head.

"You have to pay for the important things, don't you find? Sometimes we pay with money. Sometimes by giving up something we like. Sometimes with sleep. It doesn't matter so much what it is as long as it's something. Otherwise we don't appreciate it, since we rarely value what we haven't paid for."

Bill adjusted the blanket on his legs.

"I really think I'm lost," I confessed.

"So was Dante. 'I found myself in a dark wood, where the straight way was lost.' Excuse me. Would you mind getting me a drink of water? I can't seem to clear my throat," Bill said.

I opened the sliding glass door and went to the kitchen. When I returned, Bill was staring out into the morning light. I handed him the glass.

"Thank you," Bill said as he gestured to a chair. "Why don't you sit?"

I slid a chair next to Bill. Even at the height of the nineteenth floor, the sounds of the city made it difficult to hear Bill's weakened voice.

"Yes. What did you say? That you were lost? These things happen. Most of our efforts end up on the ash heap anyway. Why do you continue?" Bill turned to look at me.

"Instead of just giving up?"

Bill gestured to the city below.

"Life goes on. It doesn't require much more than finding your meals and a few comforts," Bill said. "Why torture yourself? Why not sleep a little later? Enjoy life?"

"I don't know. I'm haunted by the idea of obscurity. You know, something about 'going gentle into that good night'?"

"So you're worried about failure. What do you think your failure consists of?" Bill asked.

"Giving up," I said.

"Yes. I think that's right. But it's a question of what you give up. If you have some fixed idea about how things are supposed to turn out, it will generally bring you disappointment. It's always a surprise, really."

I looked at Bill. His skin had a parchment look that seemed almost translucent. He grimaced as a moment of pain passed through him, yet he gave no outward voice to complaint. He was dressed in a gray, collarless shirt of fine linen and a herringbone waistcoat. A gnarled wooden cane leaned against his chair. He closed his eyes briefly as if to gather his strength.

"I'd like to come back as a woodcutter, a feller of trees, if there is such a thing as another life," he said as he opened his eyes. "I spent a summer in the Northwoods when I was a young man. Working for a timber company. I was an athlete in college and quite fit at the time. I wanted to get away from the city, to pit myself against nature, you might say. It was the proud aspiration of a young man who didn't truly understand the magnitude of the forces we face. But most young men have the idea of invincibility. We have scant idea how much those forces shape us and how little we can really do.

"You might find this interesting," Bill said, handing me his cane. "No, this isn't one of the mighty timbers I felled up there in the woods. It was given to me by a Japanese friend. He was the head of a monastery I visited at the end of the war. An elegant fellow of impeccable bearing."

The cane was dark and highly polished. The wood was gnarled with knots. It had considerable heft for its size.

As I looked at the cane I realized that one of the knots held a smooth, embedded river stone. The tree that furnished the cane must have stood near the bank of a stream, where it had gradually grown around the stone.

"My Japanese friend came to see me in the hospital after my accident, before they put me back together," Bill said. "He took one look at me and said, 'Lucky, lucky man. One accident like that is worth ten thousand sittings in a monastery.'"

"Well, you can try something," Bill said after a moment. "Something out of the ordinary. To find out how you feel. After all, that's what matters. The rest is just a heap of convention, devoid of any real reason. Looking over one's shoulder for the length of a life is no way to live."

"What would you suggest?"

"I'd try to see if I could fly," Bill said.

I looked at him to see if he was smiling.

"They say the angels envy us for our possibilities," Bill continued. "I suppose that might be true in some sense. But I've long envied those with the gift of flight. These days I long to soar beyond this aching old body. But if I can speak plainly, I can see you're troubled by doubt. I think flight must take faith. Imagine the bird's first leap from the nest, trusting in instinct, propelled by a wish."

Bill gestured out over the balcony. We were nineteen floors above the city. The breeze swirled around us.

"You've suffered some setbacks. Now you think you were wrong and that sticks in your craw, right?"

"Yes, it sticks in my craw," I said.

"And you don't know what's next. Or where to turn. What else lies out there for you? You're a man. You have your needs. Why not just cut a swath through the world and leave those behind you to do the *worrying? Some prick of conscience, perhaps, holds you back. Is it real or just another weighty convention? Difficult to soar with such burdens, I should think. So I suppose you'll just have to let them go. What else is there for it?"*

"I don't know," I replied.

"I suspect you will," Bill said softly. "Though a price must be paid. That, I think, is the way of things. Or you could just forget it. Eat your three meals a day. Eke out a living. Fill a few pleasures and stop worrying about the rest. I don't think the great round world cares either way. But if I were you, I'd take the leap."

THE BEES IN WINTER

◆

Bees spend the winter huddled together in a cluster, living off the honey they have stored during the summer. The heat they give off as they metabolize the honey keeps the temperature at 90°F (32°C) in the center of the cluster. This is a challenge when the temperature outdoors can reach 20°F (28°C) below zero at times. The bees on the outside of the cluster often succumb to the cold and will drop off in their attempt to keep the bees deeper within warm. The whole point of this sacrifice is to keep the queen alive.

EACH HIVE CONTAINS A SINGLE QUEEN, and without her, the hive is finished. The queen needs to remain warm in order to remain fertile. The bees use the energy from the honey to rapidly contract their thoracic muscles without moving their wings in something akin to shivering. This muscle movement generates the heat they need to keep the queen alive until spring, when she can begin laying eggs again to rebuild the population in time for the first nectar flow.

Death in Winter
During the course of the winter, the floor of the hive becomes covered with a carpet of dead bees. During the first day of a thaw in January or February, the surviving bees will carry the corpses of their fallen sisters out the hive entrance and drop them in the snow a few feet away. The area in front of the hive

WINTER

will be littered with dead bees on those thaw days. It is a good sign when the surviving bees are strong enough to carry out this undertaking. As long as there are not too many corpses. No sign of dead bees is worse, however. In that case it means that the entire colony has succumbed to cold or disease during their confinement.

Checking the Hives

Once the bees start flying again, it's time to open the hives to see how they have fared. It used to be a rewarding and exciting task when I lifted the lids to find the hives full of bees and the queen already filling the frames with eggs and brood in preparation for the first nectar flows.

Now it's a mixed affair. Some hives still make it through with strong populations. I still find the queens here and there, scurrying around the frames. But often as not I find a diminished population barely hanging on. Or else it's silent hives with dead bees scattered in the frames or piled on the bottom board.

Last year the bears destroyed half the hives. I was able to put a few back together, but none made it through that winter.

A Poor Harvest

I had ten hives this past fall. From those ten, I only managed to harvest a meager 60 pounds (27 kg) of honey. Other years, the total has been between 600 and 800 pounds (270 and 360 kg). I had started the spring with six new hives.

The new bees were trucked up from Georgia in April and put into the hives when they arrived. Over the summer, three wild swarms had moved into the empty hives. I found a fourth swarm on a nearby tree branch and put them into the tenth hive.

By fall, the wild hives had produced a good amount. But some of the newer bees produced very little, so I had to resort to some wealth redistribution to make sure all the hives had enough honey to last them through the winter before we removed the remaining 60 pounds (27 kg) as a token of our combined efforts—bees and beekeepers.

Active or Abandoned?

I checked the hives this March. It had been a long winter, with several blizzards and a protracted spell of cold through January.

The first thing to do before opening the hives is to kneel and keep a close watch on the activity at the entrance where the bees are coming and going. You can tell quite a bit about the hive's health by the amount of activity. If there is no activity, then that is a bad sign.

On the other hand, you can also be fooled by a lot of activity. An abandoned hive can still contain stores of honey. Bees from nearby hives will do their utmost to empty it as quickly as they can, because finding honey is equivalent to finding money for us. The bees have to work hard to make honey, bringing in nectar and evaporating 80 percent of it to make one drop of honey. Finding honey is like winning the lottery.

One way to check whether a hive is abandoned and being robbed is by seeing whether bees are guarding the entrance and challenging incoming bees. The other things to look for are bees returning and entering the hive with pollen, which they carry in special pollen baskets on their hind legs. They use the pollen to feed the young bees during their larval stage. When bees are bringing in pollen it usually means there is a laying queen active inside the hive.

An Inspection Ritual

To find out which hives have survived, you need to open the hives and look inside. The first step to doing this is to prepare the smoker in the event that the bees will resent the intrusion. The theory here is that the smoke alarms the bees with the prospect of a forest fire, meaning they will have to abandon their hive and build a new one. Since the bees make wax from honey, they retreat into the hive with the first few puffs of smoke and fill themselves with honey as a feeding response to the smoke and are less likely to sting.

> A honeybee when filled with honey never
> volunteers an attack, but acts solely on the defensive.
>
> FROM LANGSTROTH'S *HIVE AND THE HONEY BEE*,
> L.L. LANGSTROTH, 1851

The next step is to remove the outer and inner covers to the hive to see inside. A first glance will tell whether the population is strong or weak. A weak population may not make it into summer, even though they have managed to survive through the harsh winter.

Next, it's time to determine whether they have an active queen. A hive without a queen or young female larva to raise a new queen is doomed. You must remove one of the frames from the outer edge of the hive to give yourself room to lift the center frames without inadvertently crushing the bees or, most importantly, harming the queen.

The Moment of Truth

I completed my observations at the hive entrances. Several hives seemed quite strong. There were bees bringing in pollen. There were bees guarding the entrances. A few other hives showed a lot of activity, but with neither guards nor bees bringing pollen in, the activity seemed suspicious. A couple of the hives were eerily quiet.

I lifted the cover of the first hive. There was a small population in the upper story. There was also a mouse in the lower story, which leapt from the hive when I lifted the box. I removed one of the side frames from the main section of the hive and then lifted out one of the center frames. I saw the queen right away. The queen is slightly longer than the female workers. The best way to spot her is to know where to look and then to search for a kind of scurrying movement that is different from that of the workers. The queen wends her way through the workers, which part as she passes, trying to escape the curious eyes of the beekeeper, it seems. One more quick glance showed some capped brood (this is when the cells of the pupae are sealed over by the bees after the eggs are laid). An active queen and capped brood are the best of signs in the early spring. The low population was of concern, since there was still all of April to get through before the nectar started flowing in earnest.

The next three hives were dead. They all had plenty of honey. I suspect the prolonged cold and their low populations going into the winter were to blame. Bees can starve and freeze, even with honey only inches away, if it's too cold for them to move off the winter cluster to feed themselves. Even though I had insulated the hives, I believe this was the case. In the end, the total was five living and five dead hives. Better than last year, but still the bees were struggling.

In anticipation of grim results, I had decided to order four more new colonies in January, to be shipped up in the first week of April.

Now it was time to clear away the remains of winter and make everything ready for the spring.

LESSONS OF THE BEES

We walked away from the Paris catacombs, all the way to the Luxembourg Gardens. There was an apiary there, with hives built to look like tiny forest huts. No one was around. We climbed the fence for a closer look. I explained how the bees wintered over inside the hive, clustered together for warmth, living off the honey they had gathered when the flowers were in bloom. It was too cold for the bees to fly, just then, so I had her put her ear to the side of the hive while I gave it a gentle tap.

"I can hear them buzzing," she said.

"They're all tucked snugly in their winter beds. Dreaming about the first flowers of spring. Printemps, *they call it."*

I looked at her kneeling beside the bees.

"I've spent all this time running away from sorrow," I said.

"Nobody runs toward it."

"And nobody outruns it, either."

"You were telling me about the bees, before we came here," she said.

"Painful lessons," I answered. "Right."

"But you kept going back."

"Trying to get it right."

"And now?"

"I'm glad you're here."

Epilogue

Roy died a few years ago. I miss him very much. Charlene and I keep in touch.

Bill Klein passed on at the age of ninety-five. We continued to work together right up to his last day.

I'm married now, with my own family. And I still get my bees from Herman West every spring.

GLOSSARY

Acarine mites Parasites that attack the trachea of young bees and lay their eggs there. Devastating to British beekeepers in the early twentieth century, acarine mites were thought to have entered the US in the 1980s.

Apiary A place where beehives are kept as managed bee colonies. The apiarist (beekeeper) determines where the apiary is located for the best yields depending on local conditions.

Beelining An ancient method used to locate wild bee colonies, using basic tracking skills and a bait to follow foraging bees back to their colony.

Bridging A term used to describe how bees cling to each other when swarming or hanging from a tree branch.

Brood chamber The nursery area of a hive where the queen lays her eggs in late winter or early spring. It is often housed in deep boxes away from the honey gathering area.

Cell capping The wax cap that is placed over the larvae cells when larvae develop into pupae. Young bees eat their way out of the cells before emerging as fully formed bees.

Colony collapse disorder When bee colonies decline significantly, leaving hives abandoned in winter. It is thought the bees may have ingested new types of pesticides (neo-nicotine insecticides) from plants which attack their central nervous system.

Crop The honey stomach of the bee. It is an organ in the abdomen where the nectar is stored temporarily to take it back to the hive. It is not part of the bee's main digestive system.

Drones Male bees that have developed from unfertilized eggs. They do no work in the colony, and their sole purpose is reproduction by mating with the queen of the hive.

Honeycomb The mass of hexagonal cells with waxy walls made by bees to store liquid honey. It is also called beeswax. To remove the liquid honey it needs to be extracted from the individual hexagonal cells.

Hybridization The search for disease-resistant species can lead to dominance of particular species. This hybridization means a lack of diversification, which can make remaining species more vulnerable.

Hypopharyngeal glands In young worker bees the glands on the head produce a secretion, which helps make up royal jelly and is fed mainly to the larvae in the hive until they are a few days old.

Larvae The stage of development after the queen has laid her eggs. For several days the larvae feed on royal jelly, before turning into pupae.

Mandibular glands In young worker bees the glands on the jaw produce a white substance that is mixed with the secretion of hypopharyngeal glands to make royal jelly. In the queen the mandibular glands are the source of the pheromones she produces.

GLOSSARY

Nectar A sugar rich liquid produced by flowers which is the main source of sugar for making honey. Often visiting bees will brush past the reproductive organs of a flower and help pollinate it.

Nosema A parasitic disease that forms in the gut of the bee and quickly spreads to the rest of the hive, shortening the life of the inhabitants, particularly over the winter months.

Pheromones Complex chemicals produced by the queen bee and used as a way to communicate and control the behavior of the bees and to attract mates within the hive.

Pollen The coarse powder produced by flowers to fertilize plants. Worker bees collect pollen from flowers in their pollen baskets and carry it back to the hives where it is fed to larvae.

Proboscis The part of bee that is used for sucking nectar, honey, and water. This long tongue moves rapidly backward and forth to collect liquid and curls up behind the head when not in use.

Propolis A resin ('bee glue') collected by bees from the buds of certain trees or sap from plants. The bees mainly use resin to strengthen their hives and keep out bad weather.

Queen bee A large, sexually mature adult female that lives in the hive and is the most important bee in the colony. Her main purpose is to lay eggs, from which all the other bees in the hive develop.

Queen excluder A device placed between the brood chamber and the honey supers to contain the queen in the brood chamber.

Round dance A particular rounded dance performed by a worker bee that indicates the location of a food source near the hive.

Royal jelly The rich, nutritious food secreted from the glands of worker bees to feed the larvae. A chosen queen is fed exclusively on royal jelly, which triggers her development into the queen of the hive.

Supers The frames of a hive that are used to store and collect surplus honey and are kept separate from the brood chamber.

Swarming When workers gather to form a new hive. If the capacity of a hive is exceeded, the workers will find a new queen and swarm to form a new colony.

Tail-waggle dance A particular figure-of-eight dance of the worker bee that involves running through a pattern of steps to indicate the direction of a food source a distance from the hive.

FURTHER READING

Buddhism & Spirituality

Nawang Gelek, Rimpoche. *Good Life, Good Death: Tibetan Wisdom on Reincarnation*, New York: Riverhead Books, 2001.

Pabongka, Rinpoche. *Liberation in the Palm of Your Hand*, Boston, MA: Wisdom Publications, 1997.

Reps, Paul (ed.). *Zen Flesh, Zen Bones: A Collection of Zen and Pre-Zen Writings*, New York: Anchor Books, 1961.

Tsongkapa. *The Principal Teachings of Buddhism*, Howell, NJ: Classics of Middle Asia, 1998.

Bees and Beekeeping

Adam, Brother. *In Search of the Best Strains of Bees*, Hamilton, IL: Dadant & Sons, 1983.

Barth, Friedrich G. *Insects and Flowers: The Biology of a Partnership*, Princeton, NJ: Princeton University Press, 1991.

Crane, Eva. *The Archeology of Beekeeping*, Ithaca, NY: Cornell University Press, 1983.

Crane, Eva. *The World History of Beekeeping and Honey Hunting*, New York: Routledge, 1999.

Frisch, Karl von. *The Dance Language and Orientation of Bees*, Cambridge, MA: Harvard University Press, 1993.

Graham, Joe (ed.). *The Hive and the Honey Bee*, Hamilton, IL: Dadant & Sons, 1992.

Langstroth, L.L. *Langstroth's Hive and the Honey-Bee*, Mineola, NY: Dover Publications, 2004.

Lindauer, Martin. *Communication among Social Bees*, Cambridge, MA: Harvard University Press, 1961.

INDEX

ALSO AVAILABLE FROM LYONS PRESS

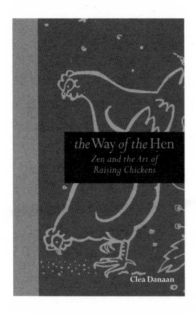

*the*Way *of the* Hen *explores the rewarding, entertaining, and enlightening art of keeping chickens in an urban or suburban garden. Chickens slow us down and ground us. This book demonstrates how raising chickens can be fitted into a busy lifestyle, and why doing so helps keep us sane and focused on the simpler joys of life.*